HUMAN ADVANTAGE

OS

OPERATING SYSTEM

COLLECTION

books about the hidden systems that shape human life.

THE HUMAN ADVANTAGE OS

The Operating System for a Human Future

Lisa J. Scott

The Human Advantage OS
The Operating System for a Human Future

This is a work of nonfiction. Names and identifying details have been modified where necessary to protect privacy.

ISBNs
Paperback: 978-1-971735-62-7
Hardcover: 978-1-971735-63-4
E-book: 978-1-971735-64-1

Published by
Lisa J. Scott Publishing
Dallas, Texas
www.lisajscottpublishing.com

LISA J. SCOTT
PUBLISHING

Printed in the United States of America

THE HUMAN ADVANTAGE OS

The Operating System for a Human Future

Lisa J. Scott

Machines are becoming more capable.

This book explores what becomes more valuable because of it.

Through the lens of artificial intelligence, automation, leadership, education, business, and everyday life, we examine the operating system beneath human flourishing.

Not the technology itself—

but the human capacities technology cannot replace.

Judgment before information.

Meaning before optimization.

Relationship before automation.

Attention before distraction.

Authenticity before simulation.

Responsibility before recommendation.

Trust before scale.

Taste before abundance.

The future does not belong to those who outrun machines.

It belongs to those who stop abandoning what machines cannot become.

The deeper question has never been whether artificial intelligence will change the world.

It is whether humanity will remember what makes us human as it does.

This is an invitation to build a life—

and a future—

around what remains irreplaceable.

For those who fear that artificial intelligence

may diminish what makes us human.

May you discover that our greatest advantages

have only become more valuable.

Preface

I did not begin this book by trying to write about the future.

I began it by noticing something small.

A humanoid robot was shown cleaning an apartment. It was not doing anything dramatic. It was not walking across a battlefield, performing surgery, composing a symphony, or solving a constitutional crisis. It was mopping a floor.

The image was almost ordinary.

That was what made it unsettling.

For years, people have spoken about artificial intelligence as if the question were mostly intellectual. Could it write? Could it code? Could it analyze? Could it pass exams? Could it reason? Could it generate images? Could it answer questions faster than a human?

Then the question moved into the room.

Could it clean?

Could it carry?

Could it move through a home?

Could it perform the physical tasks that have always anchored human life?

The robot with the mop represented something larger than a cleaning service. It suggested that automation was no longer contained inside screens. Intelligence was entering the physical world.

And when intelligence enters the physical world, the question changes.

For a long time, we assumed that human beings would always have certain protected domains. Knowledge workers believed physical labor would be automated first. Physical workers believed human judgment and creativity would remain separate. Artists believed machines could imitate but not originate. Professionals believed expertise would protect them. Parents hoped that education would prepare children for stable careers if they learned enough, performed well enough, and kept up.

But the boundaries are moving.

Artificial intelligence is making analysis, writing, research, planning, coding, and content generation easier to produce. Robotics is beginning to make physical service less dependent on human bodies. Systems that once required teams, hours, credentials, and coordination can now be initiated by one person with the right tools.

This does not mean human beings are becoming obsolete.

It means the basis of human value is changing.

That is the subject of this book.

The wrong question is: What will machines replace?

The better question is: What remains valuable when replacement becomes possible?

Every age has its own scarcity. In one age, land was scarce. In another, capital was scarce. In another, information was scarce. In another, distribution was scarce. As scarcity shifts, value shifts with it.

We are entering an age in which intelligence may become abundant.

Not wisdom. Not judgment. Not meaning. Not trust. Not responsibility.

Intelligence.

The distinction matters.

A machine may generate a hundred options. A human must still decide which one deserves to exist.

A machine may produce a thousand images. A human must still recognize which one carries meaning.

A machine may answer a question. A human must still know whether the question was worth asking.

A machine may recommend an action. A human must still be accountable for what happens next.

This book is about the human advantage that remains after intelligence becomes easy to access and labor becomes increasingly automated.

It is not an argument against technology. I use technology. I build with it. I think with it. I create with it. I have seen how powerful it can be when used as a tool for expansion rather than substitution.

But the more powerful the tools become, the more important the human operating system becomes.

Without judgment, more intelligence creates more noise.

Without taste, more creation creates more clutter.

Without trust, more information creates more suspicion.

Without meaning, more efficiency creates emptiness.

Without responsibility, more automation creates diffusion.

Without attention, more abundance creates fragmentation.

Without relationship, more connection creates loneliness.

The human advantage is not found in pretending machines are weak.

It is found in understanding where machines are strong and then asking what those strengths make more valuable.

This is the inversion.

When machines become better at producing answers, humans must become better at choosing questions.

When machines become better at generating content, humans must become better at recognizing significance.

When machines become better at performing tasks, humans must become better at designing lives, businesses, institutions, and relationships that are worth serving.

The future does not merely ask us to become more efficient.

It asks us to become more discerning.

More responsible.

More grounded.

More creative.

More relational.

More human.

This book is written for founders, parents, creators, professionals, students, teachers, and anyone trying to understand what to build in a world where old forms of advantage are weakening.

It is also written for the person who senses that keeping up is not enough.

The point is not to chase every tool, mimic every machine, or automate every corner of life.

The point is to know what should never be surrendered.

The Human Advantage Operating System begins there.

Table of Contents

Diagrams & Visual Illustrations

Introduction

The Wrong Question

The first question people usually ask is what artificial intelligence will replace.

It is an understandable question.

Replacement is visible. We can imagine a task being done by a machine. We can picture software writing a memo, drafting a contract, generating a lesson plan, designing an image, composing music, reviewing data, answering a customer, or coding a product. We can picture a robot moving through a warehouse, delivering food, driving a car, stocking shelves, folding laundry, or cleaning a floor.

The human mind reaches first for the thing it can see disappearing.

The job.

The task.

The role.

The function.

The paycheck.

So the conversation becomes anxious and repetitive.

Will AI replace lawyers?

Will AI replace writers?

Will AI replace teachers?

Will AI replace doctors?

Will AI replace artists?

Will robots replace housekeepers, drivers, caregivers, warehouse workers, restaurant workers, construction crews, and delivery workers?

Underneath the question is a quieter fear.

Will it replace me?

But replacement is not the deepest pattern.

Replacement is only the visible edge of a larger shift.

The more important question is what becomes valuable next.

Every technological era rearranges value. It does not simply remove one kind of work and leave an empty space behind. It changes what is scarce, what is abundant, what is respected, what is overlooked, what becomes easier, and what becomes harder.

When machines changed agriculture, human value did not disappear. It moved.

When factories changed production, human value did not disappear. It moved.

When computers changed office work, human value did not disappear. It moved.

When the internet made information abundant, the value of memorizing facts declined. But the value of filtering, interpreting, connecting, and applying information increased.

The pattern is not elimination.

The pattern is migration.

Value migrates.

It moves away from what technology makes abundant and toward what technology makes more necessary.

This is why the replacement question is too small.

If artificial intelligence can produce information, analysis, content, plans, summaries, images, code, and recommendations at scale, then those outputs lose some of their scarcity. They may still be useful. They may still matter. They may still be part of work. But they are no longer rare in the same way.

And when something becomes less rare, value shifts elsewhere.

The person who merely produces may become easier to replace.

The person who knows what should be produced becomes more important.

The person who merely answers may become easier to replace.

The person who knows which questions matter becomes more important.

The person who merely performs a task may become easier to replace.

The person who can design the system around the task becomes more important.

The future does not only ask, "Can a machine do this?"

It also asks, "Who decides whether this should be done?"

That is a different question.

It moves the conversation from fear to discernment.

For a long time, people treated intelligence as a form of protection. If you were smart enough, educated enough, credentialed enough, specialized enough, or technically capable enough, you were presumed to be safer. Intelligence created distance between the person and the machine.

But artificial intelligence complicates that assumption.

If intelligence itself becomes increasingly available, then intelligence alone cannot remain the highest advantage.

This does not mean intelligence becomes worthless.

It means intelligence becomes infrastructure.

Like electricity.

Like the internet.

Like search.

Like computation.

Once a capability becomes widely available, the advantage moves to how it is used, governed, trusted, directed, limited, and given meaning.

That is where the human advantage begins.

Not in having access to intelligence.

Not in producing more output.

Not in moving faster merely because machines can move faster.

The human advantage begins in the capacities that abundance does not automatically create.

Judgment.

Taste.

Trust.

Meaning.

Relationship.

Responsibility.

Attention.

Authenticity.

These are not nostalgic words. They are not soft leftovers from a more human past. They may become the central forms of value in a world where intelligence is cheap and labor is increasingly automated.

The more options a system can generate, the more judgment matters.

The more content a system can produce, the more taste matters.

The more synthetic material floods the world, the more trust matters.

The more efficient life becomes, the more meaning matters.

The more tasks are automated, the more responsibility matters.

The more noise expands, the more attention matters.

The more artificial experience becomes possible, the more reality matters.

This is the inversion.

Technology does not make the human less important.

It makes certain human capacities more important because they become the limiting factor.

A machine may generate a strategy, but it cannot fully understand the life, institution, relationship, history, risk, loyalty, fear, ambition, and consequence surrounding that strategy.

A machine may create an image, but it cannot know why one image feels alive and another feels empty unless a human has first taught the system through human perception.

A machine may recommend a decision, but it cannot carry moral responsibility for the outcome in the way a human being must.

A machine may simulate conversation, but it does not belong to a family, build a reputation over decades, keep promises through hardship, or sit beside someone in grief because love requires presence.

The question is not whether machines can do impressive things.

They can.

They will.

The question is what human beings must become when machines become impressive at scale.

That is the question this book follows.

The Human Advantage Operating System is not a manual for resisting technology. It is not a complaint about artificial intelligence. It is not a prediction that everything will collapse or a promise that everything will improve.

It is a framework for seeing the value migration already underway.

It asks what remains scarce when intelligence becomes abundant.

It asks what remains meaningful when output becomes easy.

It asks what remains human when systems become capable.

It asks how to build a life, a business, an education, a body of work, and a way of thinking that does not collapse when tools become powerful.

The wrong question is, "What will AI replace?"

The better question is, "What becomes valuable next?"

That is where the future begins.

Part I

The Value Shift

Chapter 1

The Great Inversion

Every age teaches people what to protect.

In an agricultural age, people protected land because land determined survival. In an industrial age, they protected production because factories, machinery, and scale created wealth. In an information age, they protected knowledge because information created leverage. In a credentialed age, they protected expertise because specialized knowledge opened doors that remained closed to others.

Each era developed its own form of advantage. The person who owned land possessed leverage. The person who owned machines possessed leverage. The person who controlled information possessed leverage. The person who accumulated expertise possessed leverage.

Human beings organized their lives around these scarcities. They studied for the scarce job, trained for the scarce skill, competed for the scarce credential, and protected the scarce role. Many built entire identities around capabilities that felt durable because they were difficult to acquire and difficult to replace.

Then the tools changed.

At first, the change appeared narrow. A program could write a paragraph. A model could summarize a document. A system could generate an image. A tool could draft code. A chatbot could answer a question.

Most people dismissed these developments as interesting novelties. The writing was imperfect. The images contained flaws. The reasoning was inconsistent. The answers still required review. The tools appeared useful, but not transformational.

Many people found comfort in comparing the machine to the human and noticing the machine's shortcomings. They concluded that because the technology was imperfect, the underlying system would remain unchanged.

But the deeper shift was never that machines had become perfect.

The deeper shift was that certain forms of intelligence had become available.

Not rare.

Not hidden.

Not slow.

Not expensive in the way they once were.

Available.

And availability changes value.

A capability does not need to become perfect to reshape an economy. It only needs to become good enough, inexpensive enough, accessible enough, and widespread enough to alter expectations.

The first spreadsheet did not eliminate accountants. It changed what accounting work meant. The first camera did not eliminate painters. It changed what painting meant. The first search engine did not eliminate intelligence. It changed what knowing meant.

Artificial intelligence will likely follow the same pattern.

The first generation of these tools will not eliminate human beings. They will change what human advantage means.

That is the great inversion.

For centuries, people were rewarded for capacities that machines could not easily perform. They memorized information, processed documents, produced written content, repeated procedures, followed instructions, calculated accurately, searched records, translated language, and organized routine work. These abilities created economic value because they required significant human effort.

Many of these abilities still matter.

But they no longer provide the same protection.

When a machine can perform a function that once required substantial human time, the value of that function begins to shift. The work may remain necessary, but the premium attached to performing it declines. Scarcity moves elsewhere.

This is what many conversations about artificial intelligence miss. They focus almost entirely on the visible task.

Can it write?

Can it code?

Can it teach?

Can it advise?

Can it clean?

Can it drive?

Can it negotiate?

Can it diagnose?

These questions are understandable, but they are incomplete. The more important question is not whether a machine can perform the task. The more important question is what the task becomes once a machine can perform it.

When writing becomes easy, value moves toward voice, insight, trust, and meaning. When analysis becomes easy, value moves toward judgment. When image generation becomes easy, value moves toward taste. When answers become easy, value moves toward questions. When automation becomes easy, value moves toward responsibility.

As labor becomes easier to purchase, deploy, schedule, and automate, value moves toward design, oversight, care, and human consequence.

This is the inversion.

The future does not reward people simply because they can produce. It rewards people who can discern what should be produced. It does not reward people simply because they can access intelligence. It rewards people who can govern intelligence. It does not reward people simply because they can move quickly. It rewards people who understand when speed is dangerous. It does not reward people simply because they can automate. It rewards people who know what should remain human.

For a long time, intelligence functioned as a ladder. A person could climb by knowing more, analyzing faster, solving harder problems, earning stronger credentials, and demonstrating superior mental ability. Intelligence created distance from manual labor. It created status, security, opportunity, and a sense of being less replaceable.

That hierarchy is now becoming unstable.

Not because intelligence has lost importance, but because intelligence is becoming distributed.

Capabilities that once belonged only to large organizations are increasingly moving into the hands of individuals. A small business can access strategic analysis once reserved for major consulting firms. A student can receive tutoring once available only to the wealthy. A creator can generate visual worlds without a studio. A founder can draft documents, test ideas, analyze markets, and build prototypes without hiring an entire team. A parent can create personalized lessons for a child. A lawyer can review more material in less time. A programmer can build faster than previous generations thought possible.

This is not only disruption.

It is redistribution.

Yet redistribution does not make everyone equally valuable.

Instead, it exposes a new divide.

Some people will use abundant intelligence to create more noise. Others will use it to create more clarity. Some will use automation to avoid thinking. Others will use it to free thinking. Some will generate endlessly. Others will recognize what deserves to be kept. Some will confuse speed with wisdom. Others will understand that the faster a system can move, the more important judgment becomes.

The machine does not remove the need for a human operating system.

It reveals whether one exists.

That is why this moment feels so disorienting. The old markers of advantage still look familiar, but they no longer behave the same way. A degree may still matter, but it may not protect someone from obsolete thinking. Expertise may still matter, but only if it evolves with new tools. Experience may still matter, but only when experience becomes judgment rather than habit.

The future rewards the capacities machines do not make abundant.

Machines can generate language, but they do not make wisdom abundant. Machines can produce options, but they do not make discernment abundant. Machines can imitate style, but they do not make taste abundant. Machines can simulate conversation, but they do not make trust abundant. Machines can optimize tasks, but they do not make meaning abundant. Machines can recommend decisions, but they do not make responsibility abundant. Machines can accelerate output, but they do not make attention abundant. Machines can fabricate experiences, but they do not make reality abundant.

The mistake is assuming the human advantage must be found where machines are weak.

Sometimes the human advantage becomes visible precisely where machines are strong.

The stronger machines become at generating options, the more valuable the human who can choose. The stronger machines become at producing content, the more valuable the human who can recognize significance. The stronger machines become at automating labor, the more valuable the human who can design humane systems. The stronger machines become at answering questions, the more valuable the human who can ask the right ones.

This is not a retreat from technology.

It is a reordering of value around what technology cannot supply by itself.

The great inversion is not that humans become unnecessary. The great inversion is that the most valuable human capacities may be the very capacities earlier systems treated as secondary.

Presence.

Taste.

Courage.

Care.

Discernment.

Responsibility.

Attention.

Moral imagination.

The ability to sit with complexity without rushing toward the first answer. The ability to recognize when a system is producing output but not value. The ability to protect what should not be automated simply because it can be.

These capacities have always mattered, but they were difficult to measure. They rarely fit neatly into resumes, job descriptions, performance dashboards, productivity reports, or standardized tests. They often appeared soft because older systems rewarded visible output.

But abundance changes the question.

When output is everywhere, output alone is no longer enough.

The future belongs not to the person who can produce the most, but to the person who understands what production is for.

That is the beginning of the Human Advantage Operating System.

Not fear of replacement.

Not worship of technology.

Not nostalgia for a slower world.

A clearer question:

When machines can do more, what must humans become better at being?

The answer begins with the inversion.

What was once secondary becomes central.

What was once assumed becomes scarce.

What was once invisible becomes valuable.

The human advantage moves away from performing the work and toward governing the meaning, direction, consequence, and integrity of the work.

That is where the next economy begins.

The Great Inversion

Old value stack flips: intelligence/output/labor become abundant; judgment/meaning/trust become scarce.

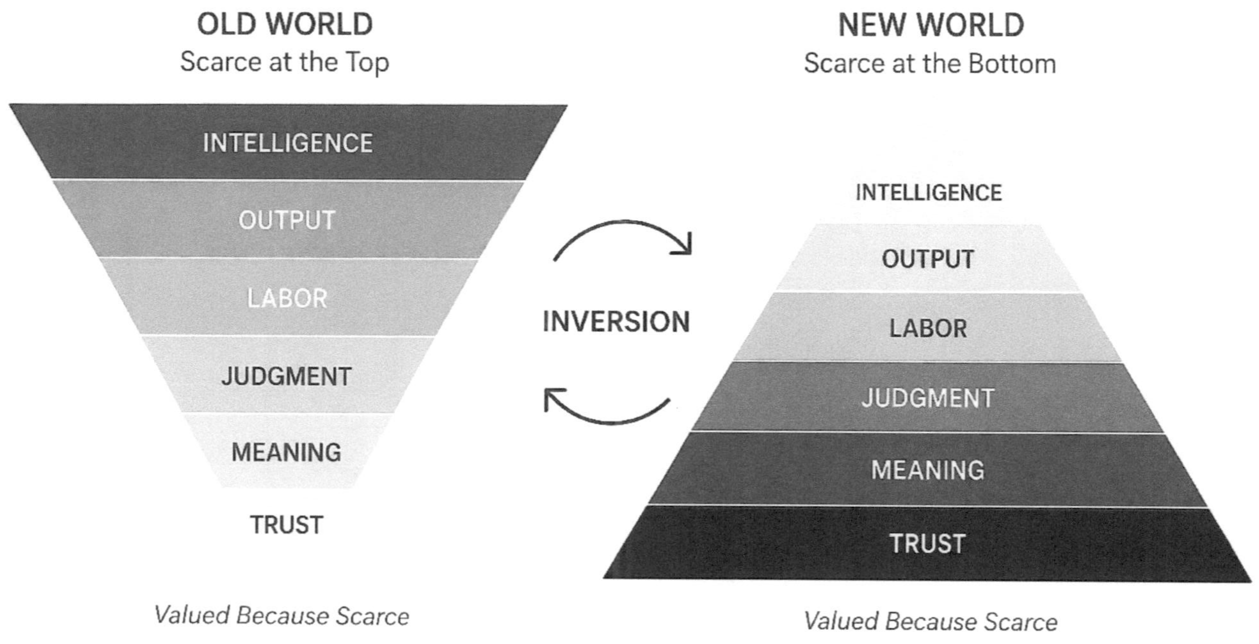

Chapter 2

When Intelligence Becomes Cheap

For most of human history, intelligence was expensive.

Not intelligence as a human trait, but intelligence as a resource.

Information was difficult to obtain. Knowledge was difficult to access. Research required effort. Analysis required time. Prediction required expertise. Creation required skill. A person might spend years accumulating knowledge that remained unavailable to most of the population.

Entire professions emerged because information was unevenly distributed. Entire institutions were built around storing, organizing, teaching, protecting, and controlling access to knowledge. If someone wanted to understand a subject deeply, they often needed to find the right teacher, purchase the right books, attend the right school, join the right organization, or spend years assembling the necessary experience.

Knowledge traveled slowly. Expertise accumulated slowly. Answers arrived slowly.

That scarcity created value.

The person who knew more possessed leverage. The person who could analyze information possessed leverage. The person who could synthesize ideas possessed leverage. In many industries, the primary advantage was simply knowing something that other people did not know.

Then the internet arrived.

Information exploded.

The transition happened gradually enough that many people failed to notice the deeper shift taking place. Information became abundant. As information became easier to access, the value of merely possessing it began to decline. At the same time, the value of navigating information began to increase.

Knowing a fact became less important than knowing where to find it. Memorization became less important than interpretation. Search replaced storage. Navigation replaced accumulation.

Many people assumed this was a temporary adjustment.

It was not.

It was the first stage of a much larger transformation.

Artificial intelligence represents the next stage.

The internet made information abundant. Artificial intelligence is making intelligence abundant.

The distinction matters.

Search engines helped people find information. Artificial intelligence helps people use it. A search engine might provide a thousand results. An artificial intelligence system can summarize those results, identify patterns, compare alternatives, explain concepts, draft conclusions, generate content, create visuals, organize arguments, and propose next steps.

What once required hours may require minutes. What once required specialists may require software. What once required teams may require a single person.

The result is not merely greater efficiency.

The result is a dramatic increase in the supply of cognitive labor.

This is where many people become uncomfortable because they instinctively understand what happens when the supply of something increases. Scarcity declines. Prices decline. Advantage shifts.

The same principle applies to intelligence.

When intelligence becomes easier to access, the scarcity that once protected certain forms of work begins to weaken. This does not mean expertise disappears. It means expertise changes.

Consider what happened to arithmetic. There was a time when performing large calculations accurately was an extraordinary skill. Accountants, engineers, scientists, and financial professionals relied on significant manual effort to perform tasks that now occur instantly.

Calculators did not eliminate mathematics.

They changed where value existed within mathematics.

The same thing happened with information. Search engines did not eliminate knowledge. They changed where value existed within knowledge.

Artificial intelligence is creating a similar shift.

The value is moving again.

For many people, this feels threatening because they built careers around capabilities that are becoming more common. Writing, research, analysis, drafting, summarization, pattern recognition, prediction, organization, translation, and content generation are not disappearing. They are becoming easier to produce.

The distinction between valuable and scarce begins to separate.

Something can remain valuable while becoming less scarce. Electricity remains valuable, but it is no longer rare. Internet access remains valuable, but it is no longer rare.

Intelligence may be moving in the same direction.

The future may not belong to those who possess intelligence.

The future may belong to those who can direct it.

This shift is already visible. A founder can perform strategic analysis that previously required consultants. A student can receive explanations once available only through tutors. A small business can create marketing content that once required an agency. A researcher can summarize large bodies of information in hours rather than weeks. A creator can generate concepts, drafts, variations, images, and experiments at a pace that would have seemed impossible only a few years ago.

The barrier to entry is falling.

The barrier to discernment is rising.

This is where many discussions about artificial intelligence become confused. People assume the question is whether machines can think.

The more important question is what happens when thinking becomes inexpensive.

Imagine a world in which nearly every person has access to extraordinary analytical capability. What becomes scarce?

Not answers. Everyone has answers.

Not content. Everyone can create content.

Not information. Everyone can access information.

Not ideas. Ideas multiply faster than they can be evaluated.

The bottleneck moves.

The constraint moves.

The scarcity moves.

And wherever scarcity moves, value follows.

In an age of abundant intelligence, judgment becomes more important. If everyone can generate options, selecting the right option becomes the challenge. If everyone can create plans, knowing which plan deserves action becomes the challenge. If everyone can produce content, knowing what is worth consuming becomes the challenge. If everyone can generate arguments, recognizing truth becomes the challenge.

Abundance creates a new problem.

Filtering.

The world has experienced this before. The internet did not fail because it lacked information. It produced too much information. Social media did not struggle because it lacked content. It produced too much content.

Artificial intelligence will not struggle because it lacks intelligence.

It may struggle because intelligence without judgment creates overwhelming complexity.

This is not primarily a technological problem.

It is a human problem.

And it may become one of the defining challenges of the century.

For generations, intelligence functioned as a sorting mechanism. Schools, corporations, professions, and institutions attempted to identify who possessed superior cognitive ability. Those individuals received opportunities, credentials, authority, and influence.

As intelligence becomes more accessible through tools, the sorting mechanism changes.

The differentiator becomes less about who can produce intelligence and more about who can use intelligence wisely.

The distinction sounds subtle.

It is not.

One person may use abundant intelligence to create confusion while another uses it to create clarity. One person may use powerful tools to amplify noise while another uses the same tools to amplify understanding. One person may use artificial intelligence to avoid thinking while another uses it to think more deeply.

The tool is identical.

The outcome is different.

The difference is the human operating system directing the tool.

This is why the future may belong to people who possess qualities that traditional intelligence tests never measured particularly well. Discernment. Taste. Wisdom. Responsibility. Curiosity. Ethics. Empathy. Judgment. The ability to recognize second-order consequences. The ability to understand context. The ability to distinguish significance from distraction. The ability to know when a technically correct answer is still the wrong answer.

Machines may become increasingly capable of generating intelligence.

Human beings remain responsible for governing its application.

That responsibility does not shrink as machines improve. It expands.

The more powerful the tool, the more important the judgment of the person using it.

A hammer requires relatively little oversight. A nuclear reactor requires a great deal.

Artificial intelligence moves humanity toward the latter category.

As capability increases, governance becomes more valuable.

The irony is that intelligence becoming cheap does not make humans less important.

It makes certain human qualities more important than they have ever been.

The people who thrive in the next era may not be the people who can produce the most intelligence. They may be the people who understand what intelligence is for.

That is the shift.

Not the disappearance of intelligence.

The disappearance of its scarcity.

And when scarcity disappears, value moves.

The rest of this book is about where it goes.

Chapter 3

When Labor Becomes Optional

For years, the conversation about automation lived almost entirely inside a screen.

The machines wrote emails, generated images, drafted reports, analyzed data, answered questions, organized information, and helped people think. Because these capabilities existed primarily in software, many people unconsciously divided the world into two categories: mental work and physical work.

The assumption seemed reasonable. Artificial intelligence would affect knowledge workers. Robots would remain expensive, limited, specialized, and uncommon. One world would change quickly while the other would change slowly.

Then the machines began leaving the screen.

Not dramatically.

Quietly.

A robot crossed a warehouse floor. A machine delivered a package. An autonomous vehicle completed another route. A robotic arm packed another shipment. A machine cleaned another building. A humanoid robot picked up another object.

None of these moments felt historic by themselves.

Most technological revolutions do not.

People often imagine revolutions as singular events—a dramatic announcement, a headline, a clear dividing line between before and after. Reality is usually less theatrical. The future tends to arrive as a collection of small observations.

A task that once required a person no longer does. A process that once demanded constant supervision becomes routine. A capability that seemed impossible last year becomes ordinary this year. A machine performs something that previously felt distinctly human.

Then another.

Then another.

Eventually the collection becomes impossible to ignore.

The significance is not simply that robots are becoming capable. The significance is that labor itself is changing categories.

For most of history, physical labor was inseparable from human bodies. If something needed to be carried, a person carried it. If something needed to be built, a person built it. If something needed to be cleaned, harvested, transported, repaired, assembled, sorted, delivered, stocked, inspected, lifted, moved, or maintained, human effort was required somewhere in the chain.

Machines amplified labor.

They did not replace most of it.

A tractor multiplied the productivity of a farmer. A forklift multiplied the productivity of a worker. A crane multiplied the productivity of a construction crew. In each case, the human remained central and the machine increased capability.

What is changing now is more fundamental.

Machines are increasingly performing labor directly rather than merely assisting it.

The distinction matters.

A software program can generate a report. A robot can now begin acting upon the physical world itself. The screen is no longer the boundary.

This changes how we think about work.

For generations, people assumed there would always be an unlimited demand for human labor because the physical world itself required human presence. Someone had to stand somewhere, lift something, carry something, watch something, clean something, or move something.

The assumption was rarely stated directly because it felt obvious.

Now that assumption is weakening.

The cost of physical automation continues to decline. The capability continues to improve. The systems continue to learn. Reliability continues to increase. Every improvement expands the range of tasks that no longer require direct human effort.

This creates understandable anxiety.

People naturally respond by asking which jobs will disappear.

The question is understandable, but it is incomplete.

The deeper question is what happens when labor itself becomes increasingly optional.

Notice the word optional.

Not eliminated.

Not extinct.

Optional.

Most technological changes do not remove all human participation. They remove the necessity of human participation.

A farmer still exists, but a tractor changed how many farmers were required. A pilot still exists, but autopilot changed what pilots spend their time doing. A warehouse worker still exists, but automation changes how many workers are needed and where their attention is directed.

The same pattern is likely to continue.

Many tasks will still be performed by humans. The difference is that they will no longer need to be.

And once necessity disappears, value moves.

This is the same pattern we explored in the previous chapter. When intelligence becomes abundant, value migrates. When labor becomes optional, value migrates again.

The future rewards less of the activity and more of the oversight. Less of the execution and more of the design. Less of the repetition and more of the judgment.

The individual who manually performs a task becomes less scarce.

The individual who understands the system surrounding the task becomes more scarce.

This shift is already visible throughout business. Many founders no longer spend most of their time doing the work. They spend their time designing the conditions under which work happens. Managers increasingly coordinate systems rather than perform every function themselves. Professionals supervise workflows that involve both humans and software.

The role changes from participant to governor.

From operator to architect.

From performer to designer.

As robotics advances, this pattern expands into the physical world.

The labor itself becomes less valuable than understanding how labor should be organized. The motion becomes less valuable than the judgment directing the motion. The action becomes less valuable than the purpose behind the action.

This creates a strange paradox.

Human beings have historically attached identity to work. People often introduce themselves through occupation.

What do you do?

The question is so common that few people examine it. Yet hidden inside the question is a powerful assumption: human value is tied to labor.

Productivity becomes identity.

Output becomes worth.

Busyness becomes virtue.

The more work a person performs, the more valuable they are assumed to be.

But what happens when productive output becomes increasingly available through machines?

What happens when effort is no longer the primary constraint?

What happens when labor itself becomes abundant?

The answer may be uncomfortable because it forces society to confront a question that has been hidden for generations.

What is a human being for?

Not what can a human produce.

Not what task can a human perform.

Not what function can a human execute.

What is a human being for?

Automation pushes that question into the open.

For centuries, necessity protected us from asking it. The farm needed labor. The factory needed labor. The office needed labor. The economy needed labor. Work provided both income and identity.

Now technology increasingly separates those concepts.

A machine may perform a task, but a machine does not possess purpose. A machine may execute labor, but it does not experience meaning. A machine may move objects through space, but it does not build a life.

This distinction becomes more important as automation expands.

The purpose of technology is not merely to eliminate work.

The purpose of technology is to reduce unnecessary effort.

These are not the same thing.

Human history has often confused labor with value. Yet some of the most valuable experiences in life are not productive at all.

A conversation with a friend.

A walk.

A book.

A meal.

A family gathering.

A piece of music.

An act of care.

A moment of reflection.

None of these activities are particularly efficient. Many produce no measurable output. Yet most people would immediately recognize them as meaningful.

This suggests something important.

Human value has never been entirely derived from labor. Labor simply occupied so much of life that people forgot to distinguish the two.

Automation forces the distinction back into view.

As machines assume more responsibility for repetitive effort, human beings become increasingly responsible for determining what that effort was meant to accomplish in the first place.

This is not a reduction of human importance.

It is a relocation of human importance.

The center of gravity moves from labor to judgment, from execution to design, from effort to meaning, and from production to purpose.

The future may not reward the people who work the hardest.

It may reward the people who understand what work is for.

That realization sits beneath every conversation about robotics. The robot cleaning the floor is not simply cleaning the floor. It is forcing a civilization to reconsider the relationship between labor, value, identity, and meaning.

The machine does not ask the question.

Its existence does.

If labor becomes increasingly optional, then the challenge is no longer how to make humans work harder.

The challenge becomes how to help humans live more deliberately.

That may be the most important transition of all.

Not the automation of labor.

The rediscovery of purpose beyond it.

The New Scarcity Map

Information → judgment, production → taste, connection → relationship, simulation → authenticity.

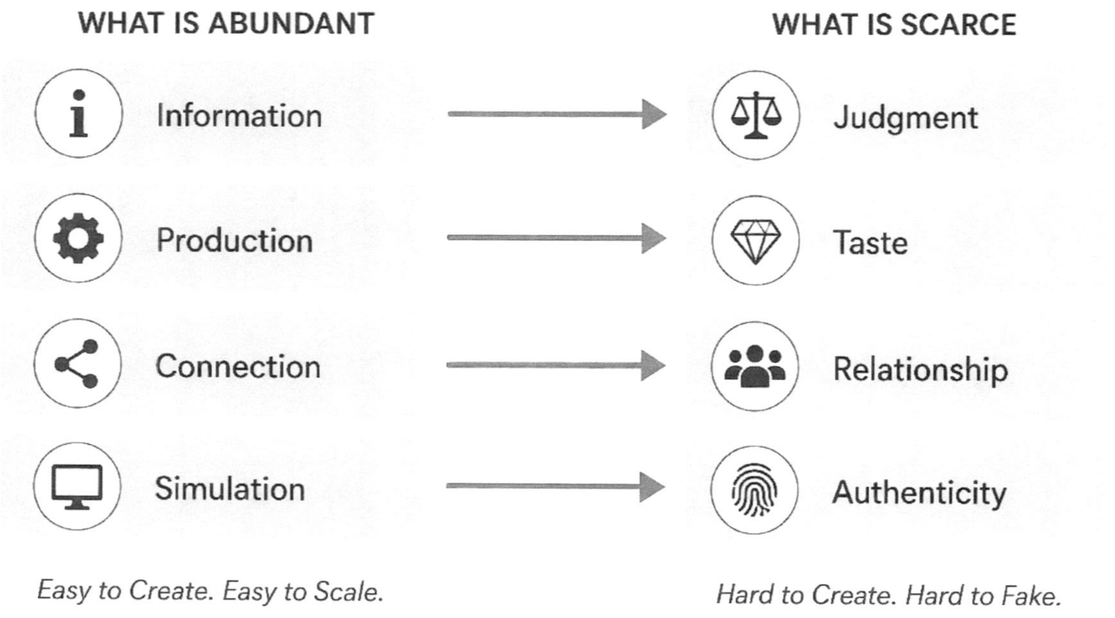

Chapter 4

The New Scarcity Map

Every economy has a scarcity map.

Most people never see it directly, but they feel it. They feel what is difficult to obtain, what is expensive, what people compete for, what institutions protect, and what society rewards. A scarcity map is more than a record of what is missing. It is a record of what a civilization has learned to value because it cannot be easily produced.

At different moments in history, different resources occupied the center of the map. In one era, land was scarce. In another, machinery. In another, capital. In another, information. In another, credentials and expertise. Each age built institutions around what it believed was valuable. Schools were created to distribute knowledge. Universities certified expertise. Corporations organized labor. Libraries preserved information. Factories multiplied production. Markets allocated resources.

These structures made sense within their own scarcity maps.

The challenge is that scarcity does not remain fixed.

When scarcity moves, institutions often continue protecting what used to be rare. They continue rewarding what used to be difficult. They continue measuring what once mattered. This is why periods of transition feel confusing. The old value system remains visible long after the underlying scarcity has begun to change.

People continue climbing ladders that no longer lead as high as they once did. They continue accumulating credentials that no longer provide the same protection. They continue producing output that no longer attracts attention. They continue gathering information in a world already drowning in it. They continue trying to prove intelligence in a world where intelligence is increasingly available on demand.

The scarcity map has shifted, but the habits of the previous age remain.

That is the condition we are entering now.

The old map rewarded processing.

The new map rewards discernment.

For generations, processing carried enormous value because processing was difficult. Someone had to read the documents, review the files, sort the data, summarize the findings, compare the options, draft the response, prepare the report, generate the content, and produce the deliverable. Processing consumed time, and time created value.

If a person could process faster, better, or more accurately than someone else, they possessed an advantage.

Today, machines are becoming increasingly good at exactly those activities. They can read faster, search faster, sort faster, summarize faster, compare faster, translate faster, predict faster, and generate faster. They do not become tired. They do not lose hours to the first draft. They do not stare at a blank page with human resistance. They can produce variations endlessly.

As processing becomes less scarce, value moves.

The question is no longer simply, "Who can process this?"

The more important question becomes, "Who can understand what matters?"

Discernment is different from processing.

Processing moves material through a system. Discernment determines what should move, what should stop, what should be ignored, what should be protected, and what should be questioned. Processing creates output. Discernment creates direction.

Processing can be accelerated.

Discernment must be cultivated.

This distinction becomes critical in an age of abundance. When there are only three options, selection is easy. When there are three thousand options, selection becomes the work. A machine may generate the three thousand. The human advantage lies in knowing which three deserve attention.

This is why output alone begins to lose power.

For a long time, output served as a proxy for value. A person wrote the report. A team produced the presentation. A company shipped the product. A creator posted the content. A worker completed the task. Visible output served as evidence that something meaningful had occurred.

But abundance weakens that signal.

When output is difficult, output impresses. When output is cheap, output must prove meaning.

A book is not valuable because words can be generated. A book is valuable because something inside it is worth saying. A strategy is not valuable because slides can be produced. A strategy is valuable because it clarifies action. A brand is not valuable because images can be made. A brand is valuable because it carries trust, coherence, and recognition. A lesson is not valuable because information can be explained. A lesson is valuable because it changes understanding.

The old scarcity map rewarded production.

The new scarcity map rewards significance.

This shift extends far beyond business.

A person can fill every hour with productivity and still feel empty. A family can become efficient and still feel disconnected. A child can complete assignments and still fail to love learning. A company can generate dashboards and still lack clarity. A creator can publish constantly and still say nothing.

More output does not automatically create more value.

More information does not automatically create more wisdom.

More connection does not automatically create more belonging.

More efficiency does not automatically create more life.

The new scarcity map exposes these realities. It reveals that many modern systems confused motion with meaning. They measured activity because activity was easy to count. They rewarded speed because speed was easy to compare. They praised productivity because productivity looked like progress.

But when machines can increase activity, speed, and productivity beyond human limits, the measurement system begins to break.

A human being cannot compete with a machine on endless output.

Nor should that become the goal.

The human advantage lies elsewhere. It lies in the ability to decide what deserves attention, recognize quality, connect action to purpose, understand consequence, and say no.

The ability to say no may become one of the most valuable capacities of the next era.

No to noise.

No to false urgency.

No to synthetic importance.

No to unnecessary automation.

No to output without value.

No to speed that damages trust.

No to efficiency that destroys meaning.

No to systems that treat human beings as obstacles rather than the reason the system exists.

This is discernment in practice.

Not passive preference.

Active governance.

The same shift changes leadership. In the old model, leaders often gained authority because they possessed more information than others. They had access to the reports, the strategy, the numbers, the institutional memory, and the private conversations.

Today, information access is becoming less differentiating.

Leadership increasingly depends on interpretation.

What does this mean?

What matters now?

What risk are we not seeing?

What pattern is repeating?

What should we stop doing?

What should remain human?

What decision are we avoiding?

These questions cannot be answered by processing alone. They require judgment inside context. They require human understanding. They require the ability to hold competing truths without collapsing into the easiest answer.

The same shift changes education.

For generations, schools trained children to retrieve, repeat, and comply. The model made sense inside an information-scarce world. Students needed knowledge, literacy, numeracy, discipline, and the ability to perform tasks on command.

Those skills still matter.

But they are no longer enough.

A child growing up in an age of abundant intelligence needs more than information. They need curiosity, discernment, creative courage, embodied experience, moral reasoning, attention, and the ability to question a machine-generated answer. They need the ability to remain with a difficult problem rather than outsourcing the first layer of thought. They need to understand the difference between assistance and dependency.

The scarcity map has shifted for children as well.

The future will not reward them merely for having answers.

It will reward them for knowing what to do with answers.

The same shift changes creators.

Content generation is becoming easier. As a result, the value of a creator can no longer rest solely on the ability to produce content. The creator must carry a world, a voice, a point of view, a lived pattern of attention, and a recognizable way of seeing.

The creator becomes valuable not because creation is difficult, but because meaningful creation remains rare.

This is why human originality may become more important, not less.

The machine can generate from what already exists.

The human can notice what has not yet been named.

The machine can imitate patterns.

The human can experience rupture, longing, grief, beauty, contradiction, and revelation from the inside.

The machine can produce language.

The human can carry meaning.

This is the new scarcity map.

The movement is consistent:

From processing to discernment.

From output to meaning.

From information to interpretation.

From speed to timing.

From productivity to purpose.

From answers to questions.

From scale to trust.

From automation to responsibility.

From synthetic abundance to authentic presence.

Once this map becomes visible, the future looks different.

It no longer appears only as a threat.

It becomes a sorting event.

Not between humans and machines, but between people who remain attached to the old scarcity map and people who learn to operate inside the new one.

Some will continue trying to prove value through volume.

Others will build value through clarity.

Some will chase every new tool.

Others will understand which tools belong inside a meaningful life.

Some will produce endlessly.

Others will discern what deserves to exist.

Some will confuse abundance with prosperity.

Others will remember that value has always belonged to what remains scarce.

The next economy begins where the old map ends.

Not with more processing.

With deeper discernment.

Not with more output.

With more meaning.

Part II

The Human Advantage Stack

The Human Advantage Stack

Layered capacities machines cannot become.

Chapter 5

Judgment

A machine can generate a thousand answers.

A human being must still decide which one matters.

This distinction may become one of the most important realities of the coming century.

For generations, intelligence was often measured by the ability to produce answers. Schools rewarded correct responses. Employers rewarded expertise. Professionals built careers around solving problems. Specialists accumulated status by possessing information that other people lacked.

The assumption was simple: the person with the answer possessed the advantage.

That assumption made sense in a world where answers were scarce. It makes far less sense in a world where answers are becoming abundant.

Today, a person can ask a question and receive multiple explanations, strategies, summaries, forecasts, recommendations, and action plans within seconds. Artificial intelligence can propose marketing campaigns, legal arguments, lesson plans, business strategies, investment theories, research summaries, product designs, and creative concepts faster than most people can fully evaluate them.

The challenge is no longer generating possibilities.

The challenge is deciding among them.

This is where judgment enters the picture.

Judgment is not the same thing as intelligence. Intelligence produces. Judgment selects. Intelligence expands possibilities. Judgment narrows them. Intelligence generates options. Judgment determines which option deserves action.

Many people assume judgment is simply another form of intelligence.

It is not.

History is filled with examples of extraordinarily intelligent people exercising terrible judgment. At the same time, some of the wisest people who have ever lived possessed only ordinary intellectual gifts.

The difference lies in what judgment requires.

Judgment lives at the intersection of knowledge, experience, context, values, consequences, uncertainty, and timing. A machine may understand a problem, but judgment understands the situation surrounding the problem. A machine may calculate outcomes, but judgment considers what those outcomes mean. A machine may identify the most efficient path, but judgment asks whether efficiency is the correct objective.

This distinction becomes increasingly important as intelligence becomes cheaper.

Imagine a leader receiving ten strategic options. A machine may rank them, model them, analyze them, simulate them, and explain them. Yet eventually someone must answer a different question:

Which path should we take?

That question cannot be solved through computation alone because the answer depends on values, risk tolerance, human consequences, long-term implications, trust, culture, purpose, and responsibility.

The decision exists inside a larger human system.

Judgment operates inside that system.

This is one reason judgment often feels slower than intelligence. Intelligence wants answers. Judgment wants understanding. Intelligence sees a solution. Judgment sees consequences. Intelligence asks whether something can be done. Judgment asks whether it should be done.

The distinction is not theoretical.

History is filled with organizations that possessed data but lacked judgment. Governments possessed intelligence but lacked judgment. Companies possessed talent but lacked judgment. Individuals possessed knowledge but lacked judgment.

The failure was rarely a lack of information.

The failure was an inability to understand what the information meant.

This pattern becomes easier to see as information expands. When information is scarce, gathering information feels like progress. When information is abundant, interpretation becomes the challenge.

Many modern problems are not caused by a lack of data. They are caused by an inability to prioritize, contextualize, and govern data. People know more than they have ever known. Yet confusion remains widespread.

Not because answers are missing.

Because judgment is scarce.

And scarcity creates value.

This is why judgment may become one of the most valuable human capacities of the next era. The machine can generate ten thousand possibilities. The human who can recognize the correct one becomes increasingly valuable. The machine can identify hundreds of opportunities. The human who understands which opportunity belongs becomes increasingly valuable.

The machine can generate arguments for every side. The human who can distinguish truth from persuasion becomes increasingly valuable. The machine can predict outcomes. The human who understands which outcomes matter becomes increasingly valuable.

In many ways, judgment functions as the governance layer above intelligence.

It sits above information, analysis, prediction, and automation. Judgment determines how those capabilities are used. Without judgment, intelligence can become dangerous.

History demonstrates this repeatedly. Technology often expands capability before wisdom catches up. The question is rarely whether something can be built. The question is whether the people building it understand the consequences.

The same principle applies to everyday life.

A parent exercises judgment when deciding how much freedom a child is ready to handle. A teacher exercises judgment when deciding when to challenge and when to encourage. A doctor exercises judgment when balancing treatment options. A founder exercises judgment when choosing growth over stability or stability over growth. A friend exercises judgment when deciding whether honesty or silence better serves a relationship.

None of these decisions can be reduced entirely to information.

Each requires context.

Each requires values.

Each requires understanding.

Each requires accepting uncertainty.

This may be the most difficult aspect of judgment. Judgment rarely operates with complete information. Intelligence prefers certainty. Judgment often functions without it.

A machine may seek the highest-probability outcome. A human being must often act before certainty arrives.

Life does not wait for perfect information.

Relationships do not wait for perfect information.

Leadership does not wait for perfect information.

The future does not wait for perfect information.

Judgment exists precisely because certainty is unavailable.

The person with judgment understands that waiting for complete clarity is itself a decision. They understand that action and inaction both carry consequences. They understand that every choice closes certain paths while opening others.

Machines may become extraordinarily capable of producing recommendations.

That does not eliminate the need for judgment.

It increases it.

Because recommendations create responsibility.

The more options a machine can generate, the more carefully humans must choose among them. The stronger the technology becomes, the more valuable judgment becomes.

This is one of the central paradoxes of the age.

Many people assume artificial intelligence reduces the need for human thinking. In reality, it may increase the importance of the highest forms of thinking. The machine can perform analysis. The human must govern analysis. The machine can generate answers. The human must decide which answer serves reality. The machine can suggest action. The human must accept responsibility for action.

This is why judgment sits at the top of the Human Advantage Stack.

Not because it is perfect.

Not because it is infallible.

But because every other capability eventually flows through it.

Trust requires judgment.

Leadership requires judgment.

Responsibility requires judgment.

Meaning requires judgment.

Relationships require judgment.

Even intelligence requires judgment.

Without judgment, intelligence becomes noise.

Without judgment, abundance becomes confusion.

Without judgment, capability outruns wisdom.

The future will contain extraordinary amounts of intelligence. That much is becoming increasingly clear.

The more important question is whether it will contain enough judgment to govern it.

The answer will not come from machines.

It will come from the humans who decide what matters.

Chapter 6

Taste

There was a time when creation itself was rare.

Writing a book required access to publishers, editors, printers, and distribution networks. Producing a film required cameras, equipment, crews, studios, and financing. Creating music required instruments, recording technology, and production resources. Design required specialized training. Art required years of technical skill.

The barriers were high.

The tools were scarce.

Creation belonged to relatively few people.

Because creation was difficult, society often treated the ability to create as the primary source of value. The creator possessed the advantage. The artist possessed the advantage. The writer possessed the advantage. The designer possessed the advantage. The musician possessed the advantage.

The challenge was producing the work.

The future introduces a different challenge.

The challenge is no longer producing the work.

The challenge is recognizing which work matters.

We are entering an age of infinite creation. A person can generate hundreds of images before lunch, thousands of words in minutes, and countless variations of logos, presentations, advertisements, videos, concepts, stories, products, and designs.

Creation itself is becoming abundant.

The result is not a shortage of content. It is an excess of content. The result is not a shortage of ideas. It is an excess of ideas. The result is not a shortage of possibilities. It is an excess of possibilities.

Whenever abundance appears, scarcity moves somewhere else.

In this case, scarcity moves toward taste.

Taste is one of the least understood human capacities. Many people treat it as preference—a matter of personal opinion, a subjective reaction, a collection of likes and dislikes.

But taste is far more significant than preference.

Taste is the ability to recognize quality before consensus arrives. It is the ability to distinguish signal from noise, identify what belongs, and recognize significance before it becomes obvious to others. Taste is pattern recognition connected to meaning.

The person with taste sees something that others overlook.

Not because they possess secret information.

Because they recognize significance.

This becomes increasingly important as creation becomes easier.

Imagine standing inside a library that contains a hundred books. Selection matters.

Now imagine standing inside a library that contains a billion books. Selection becomes everything.

The challenge shifts.

Not creation.

Recognition.

Not production.

Discernment.

Not generating.

Choosing.

This is the future many industries are approaching.

Music platforms contain millions of songs. Video platforms contain endless content. Social media generates more material in a single day than any individual could consume in multiple lifetimes. Artificial intelligence accelerates this trend by dramatically increasing the speed of creation.

The amount of content expands faster than human attention.

The amount of creation expands faster than human evaluation.

The amount of possibility expands faster than human selection.

This changes the value equation.

The creator remains important.

But the curator becomes important as well.

The editor.

The collector.

The teacher.

The reviewer.

The recommender.

The person who can point into overwhelming abundance and say:

This matters.

Pay attention to this.

Ignore the rest.

Many people underestimate the power of this ability because they assume value belongs entirely to production.

History often suggests otherwise.

A museum curator influences culture. A publisher influences culture. A teacher influences culture. A critic influences culture. An editor influences culture. A producer influences culture.

These individuals do not always create the work itself.

They recognize it.

They elevate it.

They provide context.

They help others see what deserves attention.

In an age of infinite creation, this function becomes more valuable, not less.

Taste becomes a form of navigation.

It helps people move through abundance without becoming overwhelmed.

This is why taste cannot be reduced to technical competence. Technical competence can often be taught through instruction and repetition. Taste develops differently.

Taste emerges through exposure, observation, experience, curiosity, reflection, pattern recognition, failure, and attention. A person develops taste by noticing.

The word is important.

Noticing.

Noticing what feels coherent.

Noticing what feels alive.

Noticing what endures.

Noticing what remains meaningful after novelty disappears.

Noticing what creates resonance rather than momentary stimulation.

Machines can identify patterns.

Taste requires something more complicated.

Taste requires understanding human significance.

Two images may be equally well executed. One moves people. One does not.

Two songs may follow similar structures. One survives generations. One disappears.

Two books may contain comparable information. One changes a life. One is forgotten.

Taste attempts to understand the difference.

Not mathematically.

Humanly.

This is why taste often appears mysterious. It operates in the space between analysis and intuition. The person with taste frequently recognizes quality before they can fully explain it. The explanation arrives later. Recognition comes first.

This pattern appears repeatedly throughout history.

Many innovations were initially dismissed. Many artists were initially overlooked. Many ideas were initially misunderstood. The individuals who recognized their significance early were exercising taste.

Not certainty.

Taste.

The distinction matters because taste is not prediction.

It is recognition.

The ability to see value before value becomes obvious.

This capacity becomes increasingly valuable as artificial intelligence expands creation. The future will not lack content. It will lack filters. It will not lack options. It will lack prioritization. It will not lack output. It will lack meaningful selection.

The challenge will not be finding something to watch, read, hear, purchase, learn, or create.

The challenge will be determining what deserves attention.

Taste therefore becomes a form of leadership.

A guide through abundance.

A compass inside complexity.

A way of navigating possibility.

This applies far beyond art.

Businesses require taste. A founder chooses which opportunities belong and which do not. A leader chooses which priorities deserve focus. An investor chooses which ideas merit resources. A parent chooses which values deserve reinforcement. A teacher chooses which lessons deserve emphasis.

Every meaningful act of selection involves taste.

Not merely preference.

Discernment.

The future rewards this ability because abundance creates noise. Machines can generate more possibilities than any human can evaluate. The bottleneck becomes human recognition.

Which product should exist?

Which story should be told?

Which idea deserves investment?

Which problem deserves attention?

Which opportunity belongs?

Which path aligns?

These questions sit beyond production.

They sit inside taste.

This is one reason many people misunderstand the impact of artificial intelligence on creativity. They assume that if machines can create, human creativity becomes less valuable.

The opposite may occur.

When creation becomes easy, originality becomes more valuable. When content becomes infinite, perspective becomes more valuable. When production becomes abundant, taste becomes more valuable.

The human advantage shifts from generating possibilities to recognizing significance.

The creator of the future may still create.

But creation alone will not be enough.

The creator must also possess a point of view, a worldview, a lens, a coherent pattern of attention, and a recognizable sense of what deserves existence.

This is taste.

Not decoration.

Not preference.

Not luxury.

A core human capability.

The ability to recognize what belongs in a world increasingly filled with things that can be created.

As intelligence becomes abundant and creation becomes infinite, recognition becomes power.

The person who can see what matters gains leverage. The person who can identify quality gains leverage. The person who can navigate abundance gains leverage.

Because the future will not be defined by a shortage of creation.

It will be defined by a shortage of meaningful selection.

That shortage is where taste becomes valuable.

And scarcity, as always, creates value.

Chapter 7

Trust

The problem is no longer finding information.

The problem is deciding what to believe.

For most of human history, information was scarce. News traveled slowly. Books were limited. Experts were difficult to access. Communication was constrained by geography, infrastructure, and time.

A person often trusted information because information itself was difficult to obtain. Scarcity created a natural filter. The information that arrived had usually passed through multiple layers of selection before reaching the public.

Editors selected it. Publishers selected it. Teachers selected it. Institutions selected it. Experts selected it.

The filtering process was imperfect, but it existed.

The modern world removed many of those barriers.

The internet expanded access. Social media expanded distribution. Artificial intelligence is expanding production.

The result is extraordinary abundance.

Anyone can publish. Anyone can create. Anyone can generate. Anyone can distribute. The cost of producing information continues to decline.

The cost of evaluating information does not.

This creates a problem.

When information becomes abundant, trust becomes scarce.

Scarcity creates value.

This is why trust may become one of the most important human advantages of the coming era. Not because trust is new, but because trust is becoming more necessary.

Imagine standing in a crowded room where everyone is speaking at once. The challenge is no longer finding a voice. The challenge is knowing which voice deserves attention.

This is increasingly how modern life feels.

Thousands of opinions, recommendations, headlines, experts, predictions, explanations, and pieces of content compete for attention every day. The noise expands faster than human evaluation.

At some point, people stop asking, "What information exists?"

They begin asking, "Who do I trust?"

That question changes everything.

Trust functions as a compression system. It reduces complexity. It allows human beings to navigate overwhelming environments without evaluating every piece of information independently.

No one possesses enough time to verify everything. No one can personally investigate every claim. No one can become an expert in every field.

Trust becomes the bridge.

This has always been true. A patient trusts a physician. A student trusts a teacher. A client trusts an advisor. A child trusts a parent. A citizen trusts an institution. A reader trusts an author. A friend trusts a friend.

Human civilization depends upon trust because knowledge is distributed. Trust allows people to operate beyond the limits of their own direct experience.

The difference today is that trust is becoming more visible.

In a world of synthetic abundance, trust becomes a primary filtering mechanism.

Consider what artificial intelligence makes possible. A machine can generate articles, images, videos, voices, conversations, recommendations, explanations, marketing campaigns, research summaries, and even entire identities.

The quantity is extraordinary.

The quality continues improving.

The volume becomes overwhelming.

As synthetic content expands, verification becomes increasingly difficult. Not impossible, but expensive. The more convincing a synthetic world becomes, the more valuable trusted human relationships become.

This is the paradox.

Many people assume technology reduces the importance of human connection. In reality, it may increase its value.

Trust is not simply information.

Trust is accumulated experience.

A machine can tell you something. Trust comes from observing whether that source has been right before, whether it acts consistently, whether it behaves with integrity, whether its incentives align with your interests, and whether it remains reliable under pressure.

Trust is earned through time.

Time remains difficult to automate.

A machine can generate a thousand convincing explanations. It cannot instantly generate decades of earned credibility.

This distinction becomes more important as synthetic abundance expands. The future may contain unlimited content, explanations, recommendations, perspectives, simulations, and generated expertise. Yet human beings will still ask the same question:

Who should I believe?

That question does not disappear.

It becomes more important.

This is why trusted humans become filters.

The role is subtle but powerful. A trusted person helps others navigate complexity. Not by knowing everything, but by helping determine what deserves attention.

This function appears throughout society. The trusted teacher. The trusted physician. The trusted friend. The trusted leader. The trusted parent. The trusted advisor. The trusted creator. The trusted expert.

These individuals reduce uncertainty. They provide orientation. They help people move through overwhelming environments. Their value comes less from possessing information and more from helping others interpret information.

This represents another migration of value.

In previous eras, expertise itself created leverage.

In the future, expertise combined with trust creates leverage.

Knowledge alone may no longer be sufficient. People increasingly need confidence that knowledge is being used responsibly.

Trust answers a deeper question.

Not simply, "Is this correct?"

But, "Can I rely on this source?"

The distinction matters.

Human beings do not make decisions based solely on information. They make decisions based on confidence. Confidence often rests on trust.

This is why trust may become one of the most powerful economic assets of the next era. Trust lowers friction. Trust accelerates cooperation. Trust reduces verification costs. Trust enables action.

Every transaction contains an invisible trust component. Every relationship contains an invisible trust component. Every institution contains an invisible trust component.

When trust is high, systems move efficiently. When trust collapses, systems become expensive. Everything requires verification, oversight, monitoring, and protection.

Everything slows down.

Trust is not soft.

Trust is infrastructure.

The future may make this increasingly obvious. Artificial intelligence can create extraordinary capabilities. Yet capability without trust remains fragile.

A person may have access to limitless information and still struggle to act because information does not create confidence.

Trust does.

The same principle applies to leadership. Many people assume leadership emerges from intelligence. Intelligence helps, but leadership often depends more heavily on trust.

People follow leaders they trust. Organizations function when trust exists. Teams collaborate when trust exists. Families remain resilient when trust exists.

Trust transforms information into action.

Without trust, information remains inert.

This is why trust becomes more valuable as intelligence becomes abundant. Intelligence expands options. Trust enables commitment. Intelligence generates possibilities. Trust allows people to move forward despite uncertainty. Intelligence increases capability. Trust creates coordination.

The future may therefore reward a different kind of authority.

Not authority based primarily on position.

Not authority based solely on credentials.

Authority based on earned trust.

The people who consistently demonstrate integrity, competence, reliability, honesty, and alignment may become increasingly valuable guides through complexity.

Not because they know everything.

Because people trust them enough to listen.

This applies to individuals, brands, institutions, communities, and creators. A creator who develops trust gains something more valuable than audience size.

They gain credibility.

Audience can be purchased. Attention can be borrowed. Reach can be manipulated.

Trust must be earned.

The distinction becomes more important every year.

In a world where synthetic content becomes increasingly difficult to distinguish from human content, trust becomes a signal of authenticity.

Not perfection.

Authenticity.

The trusted person does not need to be right about everything. They need to be honest about what they know, what they do not know, and how they reached their conclusions.

Trust grows through transparency, consistency, reliability, integrity, and time.

Those qualities remain difficult to manufacture.

They remain difficult to automate.

They remain difficult to scale artificially.

That is precisely why they become valuable.

The future will contain extraordinary amounts of information, intelligence, content, and simulation. What it may lack is confidence in what deserves belief.

That is where trust enters.

Not as a luxury.

Not as a soft interpersonal skill.

As one of the central filtering systems of the human future.

In a world of synthetic abundance, trusted humans become guides. They become navigators. They become filters. They become bridges between overwhelming complexity and meaningful action.

And wherever uncertainty expands, the value of trust expands with it.

Chapter 8

Meaning

A machine can answer almost any question.

That does not mean the question matters.

This distinction may become one of the defining challenges of the age of artificial intelligence.

For generations, knowledge was scarce. The challenge was obtaining answers. People invested enormous effort searching for information because information itself was difficult to access. Questions such as *How does this work? What happened? Why did this occur? What does this mean?* carried weight because the answers were difficult to find.

Answers had value because answers were scarce.

Artificial intelligence changes the equation.

Answers are becoming abundant. Explanations are becoming abundant. Recommendations are becoming abundant. Analysis is becoming abundant. Knowledge itself is becoming increasingly available.

The challenge is no longer answering questions.

The challenge is deciding which questions deserve attention.

This is where meaning enters.

Meaning does not compete with intelligence.

Meaning governs intelligence.

Intelligence tells us what can be known. Meaning helps determine what is worth knowing. Intelligence explains possibilities. Meaning determines significance. Intelligence expands options. Meaning establishes direction.

Without meaning, intelligence becomes movement without destination.

Power without purpose.

Capability without orientation.

The problem is not theoretical.

Most people have experienced it in some form. A person achieves a goal and feels unexpectedly empty. A company grows rapidly and loses its sense of identity. A student accumulates credentials and still feels uncertain. A professional climbs the ladder and discovers that the destination feels strangely disappointing.

The information was correct.

The strategy worked.

The objective was achieved.

Yet something remains unresolved.

The missing element is often meaning.

Meaning answers a different category of question. It is concerned less with *how* and more with *why*. Less with what is possible and more with what is worthwhile. Less with what can be done and more with what should be done.

This distinction becomes increasingly important as intelligence becomes more accessible.

Imagine a future in which nearly every technical question can be answered instantly. A person can ask what career to pursue, what investment to make, what diet to follow, what strategy to implement, what business to start, what city to move to, or what skills to learn. Artificial intelligence may become remarkably capable of helping people answer these questions.

Yet another question remains.

Why am I pursuing this at all?

That question belongs to meaning.

No machine can fully answer it because meaning emerges from human experience. It emerges from values, identity, purpose, belonging, love, responsibility, sacrifice, commitment, memory, hope, and loss.

Meaning exists inside a human life.

Not merely inside information.

This is why many of the most important questions people ask are not actually information problems.

A grieving person does not primarily need data. A parent does not primarily need data. A person facing mortality does not primarily need data. A person questioning purpose does not primarily need data.

Information may help.

But information alone rarely resolves the deeper question.

Meaning operates at a different level.

It organizes experience. It helps people determine what matters, prioritize competing demands, endure hardship, and decide what deserves commitment. Meaning creates coherence.

Without it, life often becomes fragmented.

One decision after another.

One optimization after another.

One achievement after another.

Yet no larger pattern connects them.

This fragmentation may become one of the hidden risks of an age defined by intelligence.

Intelligence excels at optimization.

Meaning determines what should be optimized.

Intelligence can help a person become more efficient. Meaning determines whether efficiency is serving the right objective. Intelligence can help a company grow. Meaning determines whether growth aligns with purpose. Intelligence can help a nation increase productivity. Meaning determines whether productivity serves human flourishing.

Without meaning, optimization becomes dangerous.

The system becomes increasingly effective without understanding why it exists.

Many modern institutions already struggle with this problem. Organizations become efficient while forgetting their mission. Schools become focused on metrics while forgetting learning. Healthcare systems become focused on procedures while forgetting healing. Companies become focused on growth while forgetting value. Individuals become focused on achievement while forgetting life.

The challenge is not a lack of intelligence.

The challenge is a lack of meaning.

Artificial intelligence may amplify this risk because it dramatically increases capability. The stronger the capability, the more important the purpose. A powerful tool magnifies direction. If direction is unclear, greater capability creates greater confusion.

The future therefore places increasing pressure on individuals to answer questions that cannot be outsourced.

What kind of life am I trying to build?

What relationships matter?

What responsibilities matter?

What values matter?

What trade-offs am I willing to make?

What deserves my attention?

What deserves my loyalty?

What deserves my time?

These questions do not disappear when intelligence expands.

They become more important.

Because the number of possible paths expands as well.

Abundance creates choice.

Choice creates uncertainty.

Meaning creates orientation.

This is one reason meaning often emerges through commitment rather than endless exploration. People frequently assume meaning is discovered through analysis. Sometimes it is. More often, meaning is revealed through participation.

It emerges through relationships, service, responsibility, creating something that outlives the moment, and belonging to something larger than oneself.

Meaning rarely arrives as a perfectly reasoned conclusion.

It often emerges through lived experience.

This creates another distinction between intelligence and meaning.

Intelligence can describe a relationship.

Meaning comes from being inside one.

Intelligence can explain parenthood.

Meaning comes from raising a child.

Intelligence can analyze friendship.

Meaning comes from loyalty.

Intelligence can explain love.

Meaning comes from loving.

The machine may understand the concept.

The human lives the reality.

This is why meaning remains difficult to automate.

Meaning is not information.

Meaning is interpretation.

It is the story human beings tell about experience. It is the significance assigned to events, the connection between actions and values, and the understanding that certain things matter more than others.

This hierarchy is essential.

Without it, everything competes equally for attention. Every notification, opportunity, possibility, task, and distraction appears equally important.

Meaning creates order.

Meaning allows people to distinguish between urgent and important, attractive and worthwhile, efficient and meaningful.

The future may therefore create a strange paradox.

The more intelligence becomes available, the more valuable meaning becomes.

The more questions can be answered, the more important it becomes to know which questions deserve asking. The more possibilities emerge, the more important it becomes to know which possibilities deserve pursuit. The more choices appear, the more important it becomes to know what matters.

This may explain why people often feel exhausted by abundance.

Abundance creates freedom.

Freedom creates responsibility.

Responsibility requires meaning.

Without meaning, freedom becomes overwhelming. A person drifts from one possibility to another, from one optimization to another, from one opportunity to another.

Always moving.

Rarely arriving.

Meaning provides the organizing center.

It answers the question beneath the questions.

Why this?

Why now?

Why here?

Why this life?

The machine cannot fully answer these questions because they emerge from human experience—from memory, love, grief, hope, values, relationships, and commitments freely chosen.

These realities remain profoundly human.

They cannot be reduced entirely to logic. They cannot be measured completely by data. They cannot be optimized without losing something essential.

This is why meaning belongs in the Human Advantage Stack.

Not because it opposes intelligence.

Because it directs intelligence.

Meaning functions as the compass.

Intelligence functions as the engine.

An engine without a compass can travel indefinitely without arriving. A compass without an engine cannot move.

Human flourishing requires both.

The future will almost certainly contain extraordinary intelligence.

The more important question is whether it will contain enough meaning to guide it.

Because intelligence can answer questions.

Meaning decides which questions deserve attention.

And that difference may determine whether the future becomes merely more capable or genuinely more human.

Meaning as Compass

Intelligence generates paths; meaning determines direction.

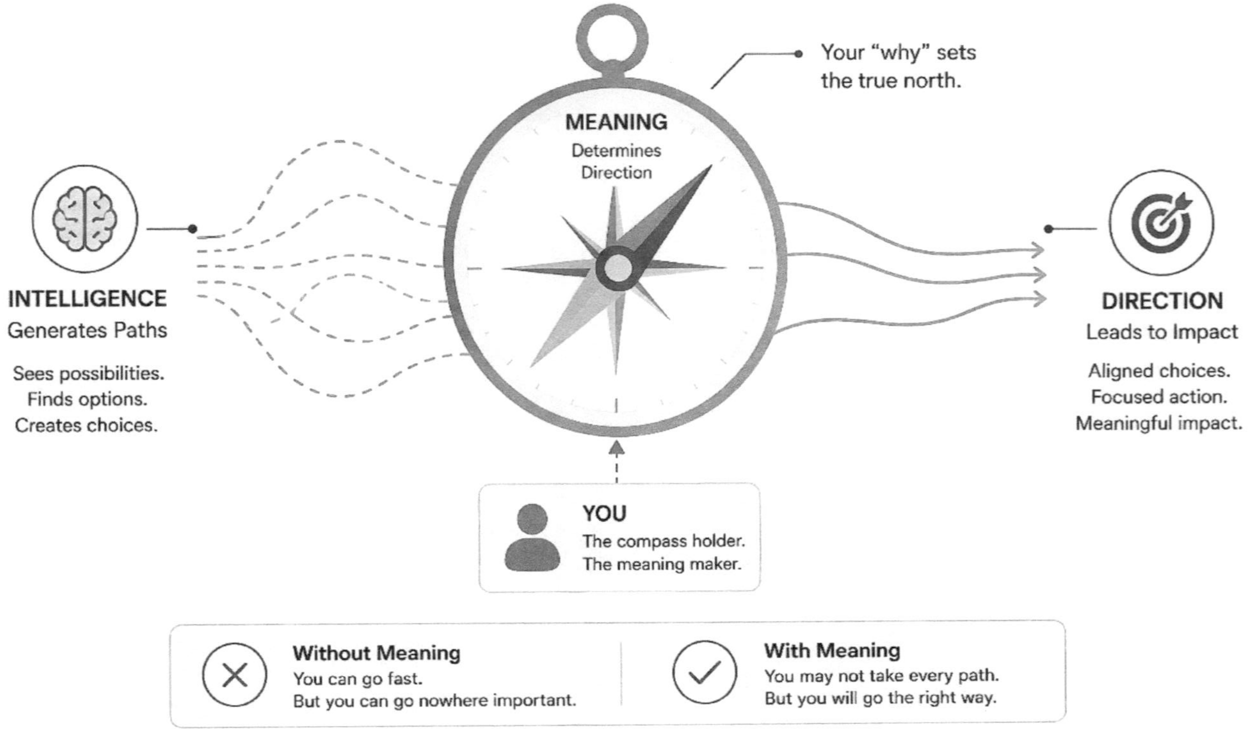

Chapter 9

Relationship

A machine can perform a task.

It cannot belong to you.

That distinction may sound sentimental until the world becomes efficient enough to reveal how much of human life was never about efficiency in the first place.

Automation is powerful because it removes effort. It completes the errand, answers the message, organizes the file, drafts the response, cleans the floor, routes the package, schedules the appointment, and performs the routine work that once consumed human hours.

This is useful.

It may even be liberating.

There are many tasks human beings do not need to suffer through in order to prove their worth. There is no moral superiority in exhaustion for its own sake. There is no sacred virtue in spending a life consumed by preventable friction.

A machine that removes unnecessary burden can return time to a person.

But time returned is not the same as life restored.

Something must fill the space.

That is where relationship enters.

For generations, modern life has often treated relationship as something left over after work was finished. Productivity came first. Obligations came first. Tasks came first. The human connections that made life meaningful were squeezed into whatever remained.

Call your mother when there is time.

See your friend when there is time.

Play with your child when there is time.

Rest beside the person you love when there is time.

Listen when there is time.

Forgive when there is time.

Be present when there is time.

The assumption was that time belonged first to production and only afterward to belonging.

Automation challenges that assumption.

If machines can perform more tasks, then human beings may be forced to confront a deeper question: What is the recovered time for?

The answer cannot simply be more automation.

A life cannot be completed like a checklist. A person cannot be loved by efficiency. A child cannot be raised through optimization. A friendship cannot be maintained through output. A family cannot be healed through productivity.

Relationship exists outside the logic of task completion.

It is not something to be finished.

It is something to be inhabited.

This is why relationship belongs in the Human Advantage Stack.

Not because machines cannot simulate conversation. They can.

Not because machines cannot respond warmly. They can.

Not because machines cannot remember preferences, produce comforting words, or imitate emotional intelligence. Increasingly, they can.

The reason is deeper.

Relationship is not merely response.

Relationship is mutual consequence.

It is the knowledge that another person's life affects yours and your life affects theirs. It is shared memory, shared risk, shared time, shared vulnerability, and shared repair.

A machine may interact.

A relationship requires belonging.

Belonging cannot be reduced to communication.

People communicate constantly and still feel alone. They exchange messages, attend meetings, post updates, react to images, follow accounts, and respond to notifications. Yet many still experience a strange absence beneath all that contact.

Connection increased.

Belonging did not.

This distinction becomes more important as technology advances. Technology can multiply interaction. It cannot guarantee relationship.

Interaction is contact.

Relationship is continuity.

Interaction can be instant.

Relationship requires duration.

Interaction can be generated.

Relationship must be lived.

The difference often appears in small moments.

A child does not only need an answer to a question. A child needs the experience of being seen.

A friend does not only need advice. A friend needs the confidence that someone remains present after the advice is given.

An aging parent does not only need assistance. An aging parent needs dignity, memory, and connection.

A partner does not only need logistics managed. A partner needs the ongoing recognition that life is being shared.

These needs do not disappear because machines improve.

If anything, they become more visible.

When machines handle more tasks, the remaining human work becomes clearer. Care, presence, loyalty, attention, repair, forgiveness, witness, and belonging begin to stand out as some of the most valuable things people provide one another.

These are not inefficient remnants of a pre-technological world.

They are central human needs.

The mistake of an efficiency-obsessed culture is assuming that anything difficult to automate must be less important.

Often the opposite is true.

The most important parts of life are frequently the least automatable.

A machine can remind you of a birthday.

It cannot carry the history behind the birthday.

A machine can write a condolence message.

It cannot grieve with the person receiving it.

A machine can help schedule dinner.

It cannot become the friendship formed across the table.

A machine can generate a bedtime story.

It cannot replace the feeling of a parent sitting close enough to be trusted.

Human beings are not only information processors.

They are relational creatures.

The nervous system understands this before the intellect does. A person can hear the same words from a stranger and from someone who loves them, and the words do not land the same way.

The meaning changes because relationship changes meaning.

This is why automation cannot fully replace belonging.

Belonging is not the content of the exchange.

It is the context in which the exchange occurs.

The same sentence carries different weight depending on who says it, when it is said, what has happened before, and what trust holds it.

Relationship creates depth.

It gives words roots.

Without relationship, communication becomes surface.

This matters for families.

It matters for work.

It matters for education.

It matters for communities.

It matters for civilization.

A workplace can automate processes and still fail because trust has collapsed. A school can use advanced tools and still fail because students do not feel known. A family can outsource tasks and still fail because no one feels emotionally present. A society can connect millions of people and still become lonely if belonging does not form.

The future will not be saved by connection alone.

It will require relationship.

This may become one of the great ironies of automation.

The more machines can do, the more valuable human presence may become.

Not presence as attendance.

Presence as participation.

The kind of presence that notices, listens, remembers, and responds to the actual person rather than merely the stated request. The kind of presence that stays when things become inconvenient, repairs after rupture, and carries responsibility beyond transaction.

Machines may become increasingly useful companions in narrow ways. They may assist, remind, entertain, teach, organize, explain, and comfort.

But they do not belong to a family in the way a person belongs.

They do not inherit memory.

They do not suffer consequence.

They do not make promises with a human life behind them.

They do not age beside us.

They do not share mortality.

Mortality gives relationship urgency.

Every human relationship is shaped by the knowledge that time is limited, even when no one says it aloud. A parent will not always be there. A child will not always be small. A friendship will not always have another decade. A conversation may be the last ordinary conversation before life changes.

This awareness creates tenderness.

It makes presence valuable.

It makes attention sacred.

It makes relationship more than exchange.

Machines do not live under that condition.

Human beings do.

That is part of the human advantage.

Not superiority.

Finitude.

Because time is limited, relationship matters.

Because people can be lost, presence matters.

Because life cannot be repeated, belonging matters.

Automation may remove many tasks from human life. It may reduce effort, increase efficiency, and expand capability.

But it cannot decide what the recovered time is for.

That decision remains human.

A person may use automation to produce more, chase more, consume more, and distract more.

Or a person may use automation to return to what was always central.

The child.

The friend.

The parent.

The partner.

The neighbor.

The community.

The conversation.

The meal.

The walk.

The shared project.

The repair.

The belonging.

This is not a retreat from the future.

It may be the point of the future.

If technology does not return human beings to one another, then its efficiency remains incomplete. Tasks may be completed, but life remains unfinished.

The Human Advantage Operating System recognizes that automation can perform tasks. It can remove burdens. It can extend capacity. It can support human life.

But it cannot replace the human need to belong.

And in a world where more tasks can be automated, belonging may become one of the clearest measures of whether progress is serving life or merely accelerating it.

Chapter 10

Responsibility

A machine can recommend.

A human being remains accountable.

This may become one of the most important distinctions of the automated age.

As artificial intelligence becomes more capable, people will increasingly turn to machines for guidance. They will ask what to write, what to buy, what to study, what to diagnose, what to invest in, what to automate, what to prioritize, what to believe, and what to do next.

The machine will answer.

Often, the answer will be useful.

Sometimes, it will be impressive.

At times, it may be more thorough, more organized, more patient, and more detailed than what a person could have produced alone.

But the answer does not carry responsibility for itself.

The recommendation may come from the machine.

The consequence enters the human world.

Someone signs the contract. Someone hires the person. Someone fires the person. Someone gives the diagnosis. Someone moves the money. Someone sends the message. Someone approves the system. Someone permits the shortcut. Someone accepts the risk.

Someone must live with what happens next.

That someone is human.

This is why responsibility belongs in the Human Advantage Stack.

It is not merely a legal concept.

It is a human condition.

Responsibility means that an outcome is not abstract. It lands somewhere. On a person. On a family. On a business. On a patient. On a client. On a child. On a community. On a future that cannot be undone.

Machines can participate in decisions.

They do not inhabit consequences in the way humans do.

A machine does not lose sleep because a recommendation damaged trust. It does not carry regret because a shortcut harmed someone. It does not repair a relationship after a careless decision. It does not explain itself to a child. It does not stand in front of a team whose confidence has been broken.

It does not answer to conscience.

This does not make machines useless.

It defines their limit.

Machines can assist with the work of thinking.

They cannot replace the moral burden of choosing.

The more powerful machines become, the more tempting it will be to hide behind them.

The report said.

The model predicted.

The system recommended.

The algorithm selected.

The automation processed.

The tool generated.

These phrases may become the new language of avoidance.

Responsibility diffuses easily inside systems.

It always has.

Large organizations have long used process to obscure accountability. A decision moves through committees, templates, policies, approvals, dashboards, and procedures until no single person appears to have chosen it.

Automation can intensify this pattern.

A person may begin to feel that if a machine suggested the decision, responsibility has shifted elsewhere.

It has not.

A recommendation is not a transfer of accountability.

It is an input.

The human remains responsible for how that input is used.

This distinction sounds obvious in theory.

In practice, it becomes much more difficult.

Machines often sound confident. They present answers with structure, authority, and polish. They compress complexity into recommendations that feel decisive. They remove the discomfort of uncertainty.

Human beings are vulnerable to this.

Certainty is attractive.

A confident answer reduces anxiety. A polished recommendation creates the feeling that someone—or something—already knows what should happen.

But confidence is not responsibility.

Fluency is not wisdom.

Clarity of presentation is not moral authority.

The human must still ask difficult questions.

Is this true?

Is this fair?

Is this complete?

Who is affected?

What is missing?

What assumptions are hidden?

What happens if this is wrong?

Who carries the cost?

These questions are the beginning of responsibility.

They slow the system down.

That slowing may be necessary.

A world that can generate answers instantly needs humans who are willing to pause before acting. Speed without responsibility becomes recklessness. Automation without accountability becomes danger. Intelligence without ownership becomes evasion.

Responsibility is the counterweight.

It reattaches consequence to decision.

This becomes especially important because many of the most consequential decisions involve people who are not present when the decision is made.

A hiring algorithm affects applicants who never meet the designers. A medical tool affects patients who never see the code. A financial model affects families whose names appear only as data. A business automation affects workers whose lives are reorganized by someone else's efficiency. A child's educational path may be shaped by systems designed far away from the child's actual mind.

In each case, the distance between decision and consequence expands.

Responsibility must travel across that distance.

If it does not, people become abstractions.

They become rows in a spreadsheet. Cases in a queue. Users in a system. Data in a model. Tasks in a workflow.

Responsibility restores the human face.

It asks who is affected by the efficiency. Who is unseen inside the optimization. Who pays for the mistake. Who benefits from the shortcut. Who remains accountable when the system does exactly what it was designed to do, but the design itself was incomplete.

This is one reason responsibility cannot be fully automated.

Responsibility requires moral imagination.

It requires the ability to picture the person affected by a decision before the consequence arrives. It requires the ability to feel the weight of choice before harm occurs. It requires understanding that a technically correct answer may still be humanly wrong.

This form of imagination is not decorative.

It is governance.

The future will need people who can govern powerful systems without surrendering moral accountability to them.

In business, this means leaders cannot simply adopt tools because they reduce cost. They must ask what kind of company the tool creates.

Does it improve service or merely reduce human contact?

Does it create clarity or hide accountability?

Does it increase capacity or erode trust?

Does it free people for better work or treat them as disposable?

Does it serve the customer or merely extract more from them?

These are responsibility questions.

In education, responsibility means adults cannot simply hand children to adaptive systems and call the result learning. They must ask what kind of learner is being formed.

Is the child becoming curious?

Capable?

Dependent?

Passive?

Resilient?

Attentive?

Does the tool deepen understanding or replace struggle too early? Does it support the child's mind or flatten it into measurable compliance?

These questions cannot be answered by performance data alone.

They require human responsibility.

In personal life, responsibility becomes equally important.

A person may use technology to optimize schedules, finances, health, communication, creativity, and productivity. But no tool can decide what kind of person that life is forming.

No tool can decide whether convenience is making someone more present or more distracted.

No tool can decide whether automation is returning energy to relationships or feeding endless output.

No tool can decide whether a faster life is a better life.

That decision remains human.

Responsibility is therefore not only about blame.

It is about authorship.

To be responsible is to remain the author of one's choices. Not the sole cause of every outcome. Not the controller of every variable. But the person willing to say:

I chose.

I approved.

I permitted.

I ignored.

I acted.

I did not act.

I must now respond.

The word responsibility contains the word *response.*

The responsible person remains able to respond.

That ability becomes more important in complex systems. When something goes wrong, the responsible person does not disappear behind the tool. They examine. They repair. They learn. They adjust the system. They tell the truth.

They accept that power and accountability must remain connected.

This may become one of the clearest dividing lines of the automated age.

Some people will use machines to avoid responsibility.

Others will use machines while deepening responsibility.

Some organizations will hide behind algorithms.

Others will build accountability around them.

Some leaders will confuse recommendation with decision.

Others will understand that the final human question remains:

Can I stand behind this?

That question cannot be outsourced.

It is the line between assistance and abdication.

A machine can help us see more. It can help us process more. It can help us generate more. It can help us predict more.

But it cannot become the conscience of a human life.

It cannot carry the burden of human consequence.

It cannot absolve us from the moral weight of choosing.

The future will contain more recommendations than any previous era. Recommendations for what to eat, buy, read, learn, automate, believe, pursue, avoid, optimize, and become.

The abundance will be useful.

It will also be dangerous if humans forget that a recommendation is not a responsibility.

The machine can recommend.

The human remains accountable.

That is not a limitation of the human advantage.

It is one of its deepest forms.

Chapter 11

Attention

The future may not be limited by intelligence.

It may be limited by attention.

For centuries, human beings struggled to gain access to information. Knowledge was scarce, communication was slow, and answers were difficult to obtain. The challenge was finding information.

Today, the challenge is often the opposite.

People struggle to escape information.

The shift is easy to overlook because abundance initially feels like progress. More information appears beneficial. More communication appears beneficial. More entertainment appears beneficial. More connection appears beneficial.

For a while, it is.

Then the volume exceeds the capacity to process it.

What was once helpful becomes overwhelming. What was once empowering becomes distracting. What was once a resource becomes a demand.

The modern world has become extraordinarily skilled at producing things that compete for attention. Notifications compete. Headlines compete. Messages compete. Platforms compete. Algorithms compete. Advertisements compete. Content creators compete.

Artificial intelligence will intensify this competition.

Not because it intends to.

Because it dramatically increases the supply of content.

A single person can now generate what once required an entire team. A company can create thousands of variations of a message. A platform can personalize content for millions of users simultaneously. An artificial intelligence system can generate articles, images, videos, conversations, lessons, summaries, recommendations, and entertainment faster than any human can consume them.

The result is predictable.

Information expands.

Attention does not.

This is the critical distinction.

Intelligence is becoming abundant. Content is becoming abundant. Creation is becoming abundant.

Attention remains finite.

There are still only twenty-four hours in a day. There are still limits to cognitive energy, concentration, and what any individual can meaningfully absorb.

The scarcity remains.

And scarcity creates value.

This is why attention may become one of the most important human advantages of the next era.

Not attention as awareness.

Attention as sustained focus.

The ability to remain with something long enough to understand it. The ability to think without interruption. The ability to resist fragmentation. The ability to direct the mind intentionally rather than surrender it to whatever arrives next.

These capacities are becoming increasingly rare.

The irony is that technology has made many things easier while making attention harder.

A person can access more information than any previous generation, yet many struggle to read a book without checking a device. A person can communicate instantly across continents, yet many struggle to remain fully present during a conversation. A person can learn nearly anything online, yet many struggle to sustain focus long enough to develop mastery.

The problem is not capability.

The problem is attention.

Attention determines what capability becomes.

The human mind does not simply absorb whatever enters it. It develops around what it repeatedly attends to.

Attention is therefore not merely a productivity tool.

It is a developmental force.

It shapes perception. It shapes memory. It shapes understanding. It shapes identity.

What a person attends to repeatedly becomes part of who they become.

This has always been true.

The difference is that previous generations lived in environments where attention was less aggressively contested.

There were distractions then, just as there are distractions now. But the modern attention economy has industrialized distraction. Entire industries compete to capture, retain, redirect, monetize, and optimize attention.

The objective is rarely malicious.

Attention is simply valuable.

Where attention goes, money often follows. Influence follows. Power follows.

The result is a world designed to interrupt.

A world optimized for engagement rather than contemplation.

A world that rewards reaction more than reflection.

This environment creates a new challenge.

The challenge is not finding something to focus on.

The challenge is protecting focus once it is found.

Artificial intelligence may amplify this challenge. Many people assume the greatest risk of AI is misinformation, unemployment, or dependency.

Those risks matter.

But another risk receives less attention.

Cognitive fragmentation.

Imagine a world where every question receives an immediate answer. Every uncertainty receives an immediate explanation. Every moment of boredom receives an immediate distraction. Every pause receives stimulation. Every difficulty receives assistance.

Convenience expands.

Patience contracts.

The human mind becomes accustomed to interruption.

The ability to remain with complexity begins to weaken.

This matters because many of the most important human activities require sustained attention. Deep learning requires attention. Creative work requires attention. Meaningful relationships require attention. Wisdom requires attention. Character formation requires attention. Understanding requires attention.

No shortcut fully replaces these processes.

A machine can summarize a book. It cannot create the experience of wrestling with an idea for hours. A machine can explain a concept. It cannot create the transformation that occurs when understanding emerges gradually. A machine can generate a plan. It cannot substitute for the attention required to execute it.

A machine can offer advice.

It cannot focus on your life for you.

This distinction becomes increasingly important as intelligence becomes more available.

The bottleneck moves.

The problem is no longer information acquisition.

The problem is information integration.

And integration requires attention.

This is why sustained focus may become a competitive advantage.

Not because it is glamorous.

Because it is increasingly rare.

The person who can think deeply gains leverage. The person who can read carefully gains leverage. The person who can listen fully gains leverage. The person who can work without constant interruption gains leverage. The person who can remain present gains leverage.

These abilities may appear ordinary.

They are not.

They are becoming scarce.

And scarcity creates value.

The importance of attention extends beyond work.

It shapes relationships as well.

One of the deepest human experiences is being genuinely attended to.

Not observed.

Not monitored.

Not managed.

Attended to.

A child experiences this when a parent listens fully. A friend experiences it when another person remains present without distraction. A partner experiences it when conversation is not competing with a device.

Attention communicates value.

It tells another person:

You matter enough for me to remain here.

The opposite is also true.

Chronic distraction communicates something else.

Not intentionally.

But powerfully.

It communicates that everything else is competing successfully for priority.

This is one reason attention may become a form of love.

Not merely a cognitive skill.

A relational one.

The future may therefore divide people in unexpected ways.

Not primarily by intelligence.

Not primarily by education.

Not primarily by access to technology.

But by the ability to govern attention.

Some people will live inside endless reaction. They will move from notification to notification, trend to trend, stimulus to stimulus, recommendation to recommendation.

Others will cultivate focus.

They will choose what receives attention. They will create boundaries. They will think before reacting. They will protect time for depth. They will resist the pressure to convert every moment into consumption.

This difference compounds over years.

Attention shapes learning.

Learning shapes capability.

Capability shapes opportunity.

Opportunity shapes life.

The future may therefore belong not to the people who have access to the most intelligence.

It may belong to the people who can remain attentive long enough to use intelligence wisely.

This is the paradox of abundance.

The more information becomes available, the more valuable focus becomes. The more content becomes available, the more valuable selection becomes. The more intelligence becomes available, the more valuable attention becomes.

Because intelligence without attention remains unused.

Knowledge without attention remains superficial.

Meaning without attention remains invisible.

Relationships without attention remain fragile.

The human advantage is not merely the ability to think.

It is the ability to direct thought.

To sustain it.

To return to it.

To protect it.

To choose what deserves it.

This is why attention belongs in the Human Advantage Stack.

It governs every other capability.

Judgment requires attention.

Taste requires attention.

Trust requires attention.

Meaning requires attention.

Relationships require attention.

Responsibility requires attention.

Without attention, the entire system fragments.

The future will contain extraordinary tools, extraordinary intelligence, extraordinary automation, and extraordinary convenience.

The rarest skill may not be producing more.

It may be remaining present.

Not for a moment.

Long enough for something meaningful to emerge.

That is attention.

And in an age of endless distraction, it may become one of the most valuable forms of human wealth.

Chapter 12

Authenticity

The more convincing imitation becomes, the more valuable reality becomes.

For much of human history, authenticity was largely assumed. A conversation happened between actual people. A photograph captured a real moment. A letter came from the person whose name appeared at the bottom. A voice belonged to the speaker. An image represented something that existed.

Authenticity was never guaranteed, but it was generally presumed.

The effort required to create convincing alternatives was substantial. Reality possessed a natural advantage because reality was easier to produce than imitation.

That advantage is changing.

Artificial intelligence can now generate images that never existed, voices that never spoke, videos that never occurred, conversations that never happened, people who never lived, and experiences that were never experienced.

Entire worlds can be created from prompts.

Entire identities can be assembled from data.

Entire narratives can be generated without direct human experience.

The technology is extraordinary.

It is also revealing.

Because the more synthetic experiences become available, the more clearly we begin to see what synthetic experiences cannot fully replace.

This chapter is not an argument against creation.

Human beings have always created stories, paintings, theater, music, literature, architecture, myths, and symbolic worlds. Imagination is not artificial. Imagination is deeply human.

The issue is not creation.

The issue is substitution.

What happens when simulation becomes easier than reality? What happens when representation becomes easier than experience? What happens when appearance becomes easier than substance?

These questions may become increasingly important as synthetic environments expand.

The challenge is not that people will lose access to reality.

The challenge is that reality may begin competing with alternatives designed to be more convenient, more personalized, more stimulating, and more immediately rewarding.

The competition is uneven.

Reality is often slow. It is messy, unpredictable, and imperfect. A garden grows gradually. A friendship develops gradually. A child matures gradually. Trust develops gradually. Mastery develops gradually. Healing develops gradually.

Reality requires patience.

Synthetic experiences can compress the timeline. They can provide immediate feedback, immediate stimulation, immediate novelty, and immediate response.

This difference creates a temptation.

Not because people dislike reality.

Because convenience is powerful.

The easier something becomes, the easier it is to choose.

Yet something interesting happens when convenience expands too far.

People begin craving what is real.

This pattern appears repeatedly throughout history. Industrial production increased the value of handmade goods. Mass communication increased the value of personal relationships. Digital music increased appreciation for live performance. Online interaction increased appreciation for in-person connection. Mass replication increased appreciation for originality.

Abundance creates its own counter-movement.

As synthetic experiences multiply, authenticity becomes scarce.

And scarcity creates value.

This is why authenticity belongs in the Human Advantage Stack.

Not because reality is always superior in every circumstance.

Because reality carries qualities that simulation cannot fully reproduce.

Consequences.

Presence.

History.

Embodiment.

Mortality.

The passage of time.

A machine can generate an image of a sunset. It cannot create the experience of standing beneath one. A machine can describe grief. It cannot lose a parent. A machine can simulate friendship. It cannot share decades of memory. A machine can write about courage. It cannot decide whether to act courageously when the outcome is uncertain.

The distinction is subtle but profound.

Authenticity is not simply factual accuracy.

Authenticity is participation in reality.

The person who climbs a mountain possesses something different from the person who viewed a simulation of one. The parent who raises a child possesses something different from the person who studied parenting. The entrepreneur who risks failure possesses something different from the person who analyzed entrepreneurship. The traveler who stands in a foreign city possesses something different from the person who viewed images online.

Information and experience are not identical.

Knowledge and participation are not identical.

Representation and reality are not identical.

The future may make these distinctions more visible.

As simulations improve, human beings may begin valuing direct experience more consciously, not less.

A meal shared around a table.

A walk without devices.

A handwritten note.

A conversation without recording.

A live performance.

A physical book.

A real garden.

A local community.

A handcrafted object.

These experiences may become more valuable precisely because they remain anchored in reality.

The value does not come from efficiency.

Often they are inefficient.

The value comes from authenticity.

They happened.

They existed.

They were lived.

Authenticity contains a form of evidence.

Not evidence for an argument.

Evidence of participation.

A life is not merely observed.

It is experienced.

This distinction becomes important because human beings derive meaning differently than machines process information.

Machines operate primarily through representation.

Human beings live through embodiment.

We occupy bodies, places, relationships, histories, communities, and moments that cannot be repeated.

A machine may understand a description of aging.

A human being experiences aging.

A machine may analyze mortality.

A human being lives with it.

A machine may summarize a marriage.

A human being inhabits one.

Reality carries weight because reality carries consequence.

The experience changes the person experiencing it.

This transformation is difficult to simulate.

It is one reason authenticity often feels recognizable even when difficult to define.

People sense the difference between something lived and something assembled.

A leader who has endured uncertainty often speaks differently than one who has merely studied it. A writer who has suffered loss often writes differently than one who only understands the concept intellectually. A teacher who has struggled often teaches differently than one who has never encountered resistance.

The authenticity comes not from perfection.

It comes from contact with reality.

This has implications for education.

A child growing up in an age of powerful simulation may need more direct experience, not less. More exploration. More experimentation. More responsibility. More observation. More encounters with the physical world.

Not because technology is harmful.

Because experience remains formative.

Children do not learn courage by reading definitions of courage. They learn courage through situations requiring it. They do not learn resilience from descriptions of resilience. They learn resilience through difficulty.

Reality teaches in ways information alone cannot.

The same principle applies to creativity.

As artificial intelligence becomes increasingly capable of generating content, audiences may become more interested in the perspective behind the content.

Who experienced this?

Who observed this?

Who lived this?

What reality gave rise to this work?

The answer becomes part of the value.

The future creator may not compete solely through production.

The creator may compete through authenticity.

Through a perspective rooted in actual experience. Through observations no machine could have independently lived. Through the particular combination of memory, attention, history, and interpretation that belongs to a human life.

This is why authenticity becomes increasingly important as synthetic abundance expands.

Not because artificial content disappears.

It will not.

Not because simulation loses value.

It will not.

But because reality becomes scarce by comparison.

And scarcity creates value.

The future may contain extraordinary virtual worlds, extraordinary simulations, and extraordinary synthetic experiences. Many will be useful. Many will be beautiful. Many will be meaningful.

Yet even in that future, people will still seek certain things.

A real friend.

A real conversation.

A real place.

A real memory.

A real achievement.

A real relationship.

A real life.

Not because simulation failed.

Because reality offers something simulation cannot fully provide.

Participation.

Consequence.

Presence.

Authenticity.

The Human Advantage Operating System recognizes that technology will continue expanding what can be created.

The more powerful creation becomes, the more important it becomes to remember what cannot be manufactured on demand.

A life lived.

A promise kept.

A friendship sustained.

A truth spoken.

A risk taken.

A moment shared.

A memory earned.

These things remain valuable because they remain real.

And when synthetic experiences multiply, reality becomes one of the rarest forms of wealth.

That rarity may become one of humanity's greatest advantages.

Part III

Work After Automation

Chapter 13

The End of Task-Based Identity

For generations, people answered a simple question in a simple way.

What do you do?

The response usually came in the form of a task.

A person taught. Sold. Managed. Coded. Drove. Cleaned. Designed. Negotiated. Diagnosed. Built. Wrote.

The answer described work.

But beneath the work sat something larger.

Identity.

Many people did not merely perform tasks. They became the person who performed those tasks. The role became part of the self. The task became part of the story. The profession became part of the identity.

This arrangement worked reasonably well for a long time because tasks were relatively stable.

A person could learn a skill, practice it for decades, and build an entire life around it. The environment changed gradually enough that identity could remain attached to function. The accountant remained an accountant. The teacher remained a teacher. The craftsman remained a craftsman. The lawyer remained a lawyer. The builder remained a builder.

The relationship between identity and work felt durable.

Automation challenges that relationship.

Not because work disappears.

Because tasks become unstable.

A task that once required years of training may suddenly require software. A process that once justified an entire department may become largely automated. A function that once occupied forty hours a week may require four.

The person remains.

The task changes.

This creates a problem for anyone whose identity was attached primarily to the task.

The more tightly identity is connected to a function, the more vulnerable that identity becomes when the function changes. This is not merely an economic challenge.

It is a psychological one.

People rarely experience disruption as a loss of process.

They experience it as a loss of self.

The question is no longer simply:

How do I perform this task?

The question becomes:

Who am I if this task no longer defines me?

This question appears throughout history whenever major technological transitions occur.

Agricultural societies experienced it. Industrial societies experienced it. Information economies experienced it.

The automation age will experience it again.

A farmer once required physical labor from sunrise to sunset. Machines changed that. A factory worker once performed tasks that became increasingly automated. Computers changed that. An office worker once spent hours organizing information. Software changed that.

Now artificial intelligence and robotics are extending the pattern.

The tasks continue shifting.

The deeper challenge is that identity often lags behind reality.

People adapt to new tools faster than they adapt to new definitions of themselves.

The spreadsheet did not simply change accounting.

It changed what it meant to be an accountant.

The internet did not simply change research.

It changed what it meant to be informed.

Artificial intelligence may not simply change work.

It may change what it means to contribute.

This possibility makes many people uncomfortable.

For centuries, work provided structure. It organized time, status, identity, and social belonging. Work answered questions that extended far beyond income.

Who am I?

What is my role?

Where do I belong?

What value do I provide?

When tasks become unstable, those answers become unstable as well.

This is why conversations about automation often trigger fear that exceeds the practical threat itself.

People are not only protecting income.

They are protecting identity.

A machine performing a task can feel strangely personal.

Not because the task was sacred.

Because the task became intertwined with self-worth.

The challenge becomes visible when people describe themselves primarily through execution.

I am the person who writes reports.

I am the person who reviews contracts.

I am the person who enters data.

I am the person who performs the analysis.

I am the person who manages the workflow.

If the task disappears, the identity becomes fragile.

Fragility emerges whenever value depends entirely on a specific function.

The automation age exposes this fragility.

The person who defines themselves by tasks becomes vulnerable to every technological advance. The person who defines themselves by deeper capacities becomes more resilient.

This distinction matters.

Tasks change.

Capabilities endure longer.

A report writer may struggle when reports become automated. A person whose deeper capability is judgment remains valuable.

A data processor may struggle when processing becomes automated. A person whose deeper capability is interpretation remains valuable.

A scheduler may struggle when scheduling becomes automated. A person whose deeper capability is coordination remains valuable.

A content producer may struggle when content generation becomes automated. A person whose deeper capability is meaning-making remains valuable.

The future rewards the underlying human capability rather than the visible task.

This shift requires a different way of thinking about work.

Many people ask:

What job will survive?

The better question may be:

What human capability does this job express?

A teacher's deeper value is not delivering information.

A teacher's deeper value is developing human beings.

A physician's deeper value is not simply providing medical facts.

A physician's deeper value is exercising judgment and care under uncertainty.

A leader's deeper value is not managing tasks.

A leader's deeper value is creating direction, trust, and coordination.

A parent's deeper value is not completing household functions.

A parent's deeper value is nurturing a life.

The task is visible.

The capability underneath the task is what endures.

This distinction becomes increasingly important because technology often removes the surface layer first.

The visible process changes.

The deeper human need remains.

People still need guidance.

They still need trust.

They still need meaning.

They still need belonging.

They still need judgment.

They still need accountability.

They still need care.

The delivery mechanism changes.

The underlying need persists.

This realization creates a more stable foundation for identity.

A person no longer defines themselves by a specific activity.

They define themselves by the value they create.

The distinction sounds subtle.

It is profound.

A person who identifies as a contract reviewer may feel threatened by automation. A person who identifies as someone who reduces uncertainty in complex transactions sees automation differently.

The capability survives even if the process changes.

A person who identifies as a content creator may feel threatened by artificial intelligence. A person who identifies as someone who helps others see clearly possesses a far more durable foundation.

The medium may change.

The purpose remains.

This is the beginning of post-task identity.

The person is no longer anchored to the activity.

The person is anchored to the contribution.

The future may increasingly require this shift because automation does not simply remove tasks.

It continuously rearranges them.

The half-life of specific skills may shorten. Tools may evolve rapidly. Entire workflows may change within a decade.

Identity cannot afford to remain attached to every changing process.

Something deeper is required.

The irony is that automation may eventually force people to ask questions they should have asked all along.

What is my actual value?

What am I helping accomplish?

What remains if the tool changes?

What remains if the process changes?

What remains if the title changes?

What remains if the task disappears?

The answers often reveal something important.

The most valuable part was never the task itself.

The task was simply the vehicle.

The deeper contribution existed underneath it.

The future belongs to people who understand the difference.

Not because they are immune to change.

Because they are not defined by the changing thing.

A machine may perform the task. A machine may accelerate the process. A machine may reduce the labor. A machine may alter the workflow.

But the human capability underneath the task remains relevant.

The person who knows how to create trust remains valuable.

The person who knows how to exercise judgment remains valuable.

The person who knows how to build relationships remains valuable.

The person who knows how to recognize meaning remains valuable.

The person who knows how to assume responsibility remains valuable.

The person who knows how to direct attention remains valuable.

These capabilities belong to the human operating system.

They are more durable than any individual task.

This is why automation represents more than a technological transition.

It is an identity transition.

The end of task-based identity is not the end of human value.

It is an invitation to discover where human value was located all along.

Not in the task.

In the person.

Not in the function.

In the capability.

Not in the activity.

In the contribution.

The future may automate many things.

The question is whether people can separate who they are from what they do long enough to recognize what remains.

That recognition is the beginning of resilience.

And resilience may become one of the defining strengths of the human future.

Chapter 14

From Performer to Governor

For a long time, work meant performance.

A person performed the task. They answered the phone, typed the report, reviewed the file, drove the truck, cleaned the room, prepared the lesson, processed the invoice, managed the schedule, or drafted the document.

The worker stood close to the work.

Their value was visible because their effort was visible.

If the task happened, someone had performed it. If the output existed, someone had produced it. Work was measured through direct participation.

Then tools began widening the distance between the human and the task.

A spreadsheet could perform calculations. A database could organize records. A search engine could retrieve information. Software could route messages. Automation could complete routine steps. Artificial intelligence could draft, summarize, compare, analyze, generate, and recommend.

Eventually robotics began carrying the shift into the physical world.

The person did not disappear.

But the person's role changed.

Less direct execution.

More direction.

Less manual performance.

More oversight.

Less doing every step.

More designing the system that performs the steps.

This may be one of the most important transitions in work after automation.

The human moves from performer to governor.

A performer completes the work.

A governor directs the system through which the work is completed.

The distinction is subtle at first.

A performer asks, *What do I need to do?*

A governor asks, *What should happen, why should it happen, who should do it, under what conditions, with what safeguards, and toward what end?*

A performer focuses primarily on execution.

A governor focuses on design, judgment, responsibility, and consequence.

This shift is already visible across many professions.

The writer no longer only writes. The writer prompts, edits, selects, refines, rejects, and gives the work a voice.

The lawyer no longer only reads every document from beginning to end. The lawyer designs review systems, identifies risk, interprets significance, and decides what deserves escalation.

The teacher no longer only delivers information. The teacher curates tools, observes understanding, guides attention, and protects the human development of the student.

The founder no longer performs every function. The founder designs the operating system, establishes standards, allocates authority, and decides what the company is becoming.

The manager no longer merely supervises people. The manager increasingly supervises workflows, tools, incentives, information, exceptions, and trust.

The role changes because the work changes.

Automation removes some of the friction that once required direct human effort. But it also creates new layers of responsibility.

Someone must decide which tool belongs inside the system.

Someone must know when the tool is wrong.

Someone must understand the limits.

Someone must protect the standard.

Someone must notice unintended consequences.

Someone must remain accountable when the system produces harm.

The human role does not vanish.

It rises one layer higher.

This is the governance layer.

The governance layer is where purpose meets process. It is where judgment directs capability. It is where human values shape automated systems. It is where efficiency is prevented from becoming careless.

A system without governance may produce more work faster.

But faster is not the same as better.

A machine can generate a contract summary. A human must know whether the summary missed the provision that matters.

A machine can create a marketing campaign. A human must know whether the campaign fits the brand.

A machine can recommend a hiring decision. A human must know whether the recommendation is fair, complete, and aligned with organizational values.

A machine can generate a lesson. A human must know whether the child actually understands.

The work may move faster.

The need for human judgment remains.

In some ways, the need increases.

A person performing a task directly experiences the work at close range. They see the details. They feel the friction. They notice what does not fit.

Automation creates distance.

Distance creates risk.

The person directing the system may no longer see every detail unless the system is intentionally designed to surface what matters.

This is why governance is not passive oversight.

It is active design.

A good governor of work understands flow.

Where does information enter?

Where does judgment occur?

Where can error hide?

Where can trust weaken?

Where do exceptions appear?

Where does the human need to intervene?

Where should automation stop?

These questions matter because automated systems tend to expand.

Once a tool proves useful, people naturally begin applying it more broadly. The temptation is understandable. If automation can save time in one area, perhaps it can save time everywhere.

But not every process should be automated in the same way.

Some work is routine.

Some work is relational.

Some work is sensitive.

Some work is developmental.

Some work carries moral consequence.

The governor understands the difference.

This may become one of the defining skills of future work: knowing which layer belongs to the machine and which layer must remain human.

A machine may draft.

A human approves.

A machine may categorize.

A human interprets.

A machine may suggest.

A human decides.

A machine may monitor.

A human responds.

A machine may accelerate.

A human governs direction.

The transition from performer to governor therefore requires a new kind of literacy.

Not merely technical literacy.

Systems literacy.

The ability to understand how tools, people, incentives, information, risk, and outcomes interact.

A person who lacks systems literacy may respond to automation in one of two ways.

They may overtrust it.

Or they may reject it entirely.

Both responses are incomplete.

Overtrust leads to abdication.

Rejection leads to irrelevance.

Governance requires a third position.

Use the tool.

Understand the tool.

Question the tool.

Limit the tool.

Improve the tool.

Remain responsible for the tool.

This is the mature human role after automation.

It is not anti-technology.

It is not blind adoption.

It is directed integration.

The performer asks the tool to help complete the task.

The governor asks whether the task belongs inside the system at all.

This is why governance is more strategic than performance.

A performer may become highly skilled at a task that later becomes automated.

A governor develops the capacity to redesign the work as conditions change.

This creates resilience.

When the tool changes, the governor adapts.

When the process changes, the governor reconfigures.

When the environment changes, the governor updates the system.

Identity is no longer tied to one form of execution.

It is tied to the ability to direct work intelligently.

This does not mean direct performance becomes worthless.

Some work will always involve embodied skill, craft, care, and presence.

A musician still performs.

A surgeon still operates.

A teacher still teaches.

A parent still shows up.

An artist still creates.

But even in these domains, the human increasingly decides how technology belongs.

The musician may use tools without losing voice.

The surgeon may use robotics without surrendering judgment.

The teacher may use artificial intelligence without abandoning the child.

The artist may use generation without losing authorship.

The question is no longer whether the human performs or the machine performs.

The question is how the human governs the relationship between capability and meaning.

This is where many organizations will struggle.

They will adopt tools without building governance. They will automate tasks without rethinking responsibility. They will reduce headcount without redesigning trust. They will increase output without clarifying purpose.

They will assume capability is the same as progress.

It is not.

Progress requires governance.

The automated organization must still answer human questions.

What standard are we protecting?

What errors can we tolerate?

What risks must be escalated?

What should never be automated without review?

Who is accountable?

Where does human judgment enter?

What kind of work are we freeing people to do?

These questions determine whether automation strengthens the system or weakens it.

The same principle applies to individuals.

A person can use powerful tools to become more scattered or more focused. More reactive or more deliberate. More dependent or more capable. More efficient or more meaningful.

The tool does not decide.

The human operating system decides.

This is the heart of the transition from performer to governor.

The human being is no longer valuable only because they can execute the step.

The human being becomes valuable because they can decide how steps should be arranged, what purpose they serve, and where the system requires judgment.

This is not a lesser role.

It is a higher one.

But it requires development. It requires attention. It requires responsibility. It requires the willingness to remain accountable for systems that can act faster than any individual human could.

The future of work may therefore reward people who can stand above the workflow without losing sight of the human beings inside it.

People who can use automation without becoming automated.

People who can design systems without surrendering conscience.

People who can direct intelligence without confusing it for wisdom.

People who understand that the work is not only what gets done.

The work is also the system that decides what gets done.

The shift from performer to governor is not the end of human work.

It is the relocation of human work into the layer where meaning, judgment, responsibility, and design meet.

That may be where the deepest human advantage begins.

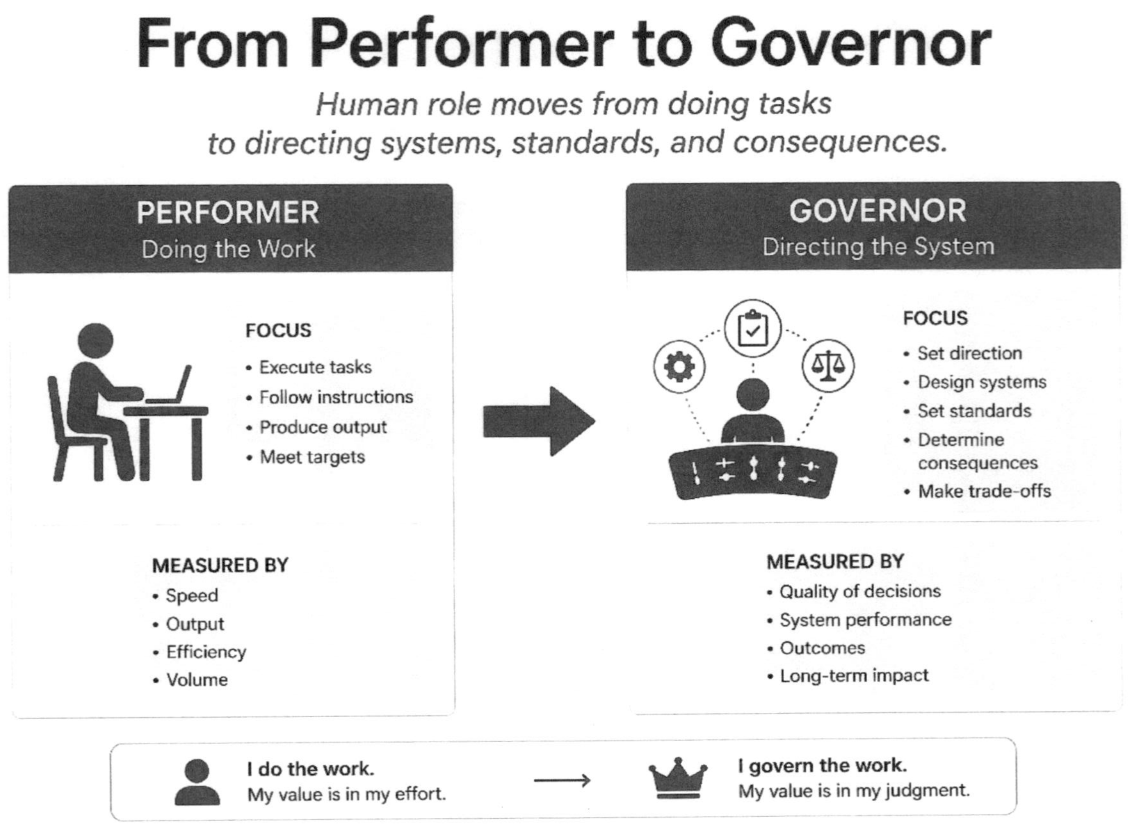
From Performer to Governor
Human role moves from doing tasks
to directing systems, standards, and consequences.
PERFORMER
Doing the Work
FOCUS
• Execute tasks
• Follow instructions
• Produce output
• Meet targets
MEASURED BY
• Speed
• Output
• Efficiency
• Volume
GOVERNOR
Directing the System
FOCUS
• Set direction
• Design systems
• Set standards
• Determine consequences
• Make trade-offs
MEASURED BY
• Quality of decisions
• System performance
• Outcomes
• Long-term impact
I do the work.
My value is in my effort.
I govern the work.
My value is in my judgment.

Chapter 15

The Judgment Economy

The future may not reward the person who sees the most.

It may reward the person who knows what does not deserve to be seen.

This idea can be difficult to understand inside a culture that has long treated more as better. More information. More data. More options. More reports. More messages. More meetings. More content. More visibility. More analysis.

Modern systems often assume that if something can be captured, it should be captured. If it can be measured, it should be measured. If it can be reported, it should be reported. If it can be generated, it should be generated.

For a long time, that logic appeared reasonable.

Then abundance arrived.

Abundance creates a different problem.

Too much eventually becomes its own form of blindness.

A person can drown in information while still missing the truth. A company can build dashboards while still fail to see the risk. A leader can receive constant updates while still misunderstand the system. A worker can process messages all day and never touch what actually matters.

This is the beginning of the judgment economy.

The most valuable workers are no longer simply the people who can process the most material. They are the people who know what to ignore, what to trust, and what to escalate.

Those three abilities may become more important than speed itself.

Ignoring sounds careless until the world becomes too full to process completely.

Then ignoring becomes a discipline.

A person who cannot ignore is controlled by whatever arrives. Every message becomes urgent. Every task becomes equal. Every opinion becomes a demand. Every metric becomes meaningful. Every generated option receives attention simply because it exists.

This is not openness.

It is vulnerability.

A system without filters can be overwhelmed by noise.

The same is true of a person.

Judgment begins with the ability to distinguish between signal and demand.

Some things are loud but unimportant. Some things are quiet but consequential. Some things appear urgent because they arrived recently. Some things are important because they alter the future.

The worker with judgment understands the difference.

They do not treat every input equally. They understand that attention is a limited resource and they protect it accordingly.

This ability becomes especially valuable when artificial intelligence increases the volume of possible work. A machine can generate more drafts than a human can read, more summaries than a team can verify, more ideas than a company can execute, more risks than a leader can address, and more recommendations than a system can responsibly adopt.

The bottleneck becomes human judgment.

What should be ignored?

What should be trusted?

What should be escalated?

These questions form the operating logic of the judgment economy.

The ability to ignore is not the ability to dismiss everything.

It is the ability to know what does not belong in the decision.

This requires context.

A junior worker may process everything because they do not yet know what matters. An experienced worker begins recognizing patterns. They understand that some issues are routine, some are noise, some are performative urgency, some are recycled concerns, and some are simply not worth the cost of attention.

This recognition saves time.

More importantly, it protects quality.

Attention spent on the wrong thing is not neutral. It steals from the right thing. Every unnecessary escalation delays a necessary one. Every irrelevant detail obscures the relevant detail. Every low-value task consumes energy that could have served judgment.

The judgment economy therefore rewards workers who can reduce cognitive clutter for the system.

They become filters.

Not passive filters.

Intelligent filters.

They absorb complexity and return clarity.

This kind of worker becomes invaluable because every organization produces noise. Emails multiply. Meetings multiply. Requests multiply. Exceptions multiply. Tools multiply. Reports multiply.

Without judgment, the organization becomes a fog machine.

Everything appears active.

Nothing becomes clear.

The second skill is knowing what to trust.

This is more difficult than it sounds.

In the past, trust often attached itself to position. People trusted the official report, the approved source, the credentialed expert, the senior leader, or the established process.

Those sources still matter.

But the modern environment is more complicated.

Information can appear polished and still be wrong. A summary can appear confident and still miss the key issue. A model can appear precise and still rest on flawed assumptions. A person can speak with authority and still misunderstand the problem.

Even a tool can produce a clean answer while hallucinating structure where no certainty exists.

The worker with judgment does not trust presentation alone.

They examine source, context, incentives, reliability, and consequence.

Where did this come from?

Who benefits if it is believed?

What does it omit?

Has this source been reliable before?

Is the confidence proportional to the evidence?

What would change if this were wrong?

These questions create a trust filter.

Trust is not blind acceptance.

It is earned reliance.

In the judgment economy, trusted workers know how to assign confidence appropriately. They do not treat all information as equally credible. They do not treat all tools as equally reliable. They do not treat all experts as equally relevant. They do not treat all data as equally meaningful.

They understand that different sources deserve different levels of reliance.

This ability becomes essential when human-generated and machine-generated information mix together.

A worker may receive an AI summary, a client statement, a legacy document, a dashboard, a meeting note, a legal requirement, a sales promise, and a manager's interpretation of events.

The task is not merely to collect them.

The task is to know how much weight each deserves.

That is judgment.

The third skill is knowing what to escalate.

Escalation is one of the most misunderstood functions inside organizations.

Some people escalate everything because they fear responsibility. Others escalate too little because they fear appearing incapable.

Both approaches create risk.

Escalating everything burdens the system.

Escalating nothing hides danger.

Judgment lives between the two.

The worker with judgment understands thresholds.

They know when something can be handled locally and when it requires broader attention. They recognize when an issue affects risk, trust, money, timing, people, standards, or the future itself.

Escalation is not simply passing work upward.

It is surfacing significance.

A strong escalation does not say, *"I do not know what to do."*

It says, *"This matters enough that the right level of judgment must enter."*

That distinction changes the value of the worker.

The person who knows what to escalate protects the organization from hidden risk. They also protect leaders from unnecessary noise.

They become a stabilizing layer inside the system.

This becomes especially important in automated environments.

Automation handles routine flow.

But exceptions remain.

Edge cases remain.

Ambiguity remains.

Human consequences remain.

A machine may process ninety percent of a workflow effectively. The remaining ten percent may carry most of the risk.

The worker with judgment notices when the ordinary pattern has broken. They see the anomaly. They recognize the mismatch. They sense that something does not fit.

They do not keep the machine moving simply because the machine is moving.

They interrupt when interruption is necessary.

This may become one of the most valuable human roles in automated systems.

Exception recognition.

Risk recognition.

Context recognition.

Human consequence recognition.

The machine processes the pattern.

The human notices when the pattern is misleading.

The judgment economy therefore does not eliminate workers.

It changes what makes workers valuable.

The valuable worker is not merely fast. The valuable worker is not merely compliant. The valuable worker is not merely available.

The valuable worker protects the system from blindness.

They know what to ignore.

They know what to trust.

They know what to escalate.

These abilities require more than technical training.

They require exposure, experience, context, and courage.

They require the willingness to say, *"This is not important,"* when everyone else is reacting.

They require the willingness to say, *"I do not trust this yet,"* when everyone else is impressed.

They require the willingness to say, *"This needs attention,"* when everyone else wants speed.

Judgment often creates friction.

That is why it can be difficult to value.

A person with judgment may slow a process down. They may question a conclusion. They may reject a shortcut. They may refuse to act on incomplete information.

To a system obsessed with speed, this can look inefficient.

But in complex environments, good friction prevents bad failure.

The future will need people who understand this.

Not people who resist every tool.

Not people who worship every tool.

People who can govern flow.

People who know when to let the system run and when to intervene.

People who understand that some of the most important work is the work that prevents the wrong thing from happening.

This work is difficult to measure.

A crisis avoided rarely produces a visible achievement. A bad decision stopped may never appear in a performance report. A risk escalated early may be forgotten once it is resolved.

Yet these acts create enormous value.

They preserve trust.

They prevent waste.

They protect relationships.

They reduce harm.

They keep systems from collapsing under their own speed.

The judgment economy will eventually learn to value this more explicitly because abundance makes it necessary.

When intelligence is scarce, the person who can produce an answer is valuable.

When intelligence is abundant, the person who can evaluate the answer becomes valuable.

When content is scarce, the person who can create content becomes valuable.

When content is abundant, the person who can identify what deserves attention becomes valuable.

When labor is scarce, the person who can perform the task becomes valuable.

When automation expands, the person who can govern the task becomes valuable.

Judgment is the throughline.

It is the human capacity that turns abundance into usefulness.

Without judgment, abundance becomes overload.

With judgment, abundance becomes leverage.

The future will contain more information, more analysis, more recommendations, more automation, and more synthetic output than any previous age.

The valuable worker will not be the one who tries to absorb all of it.

The valuable worker will be the one who knows how to move through it without losing sight of what matters.

What to ignore.

What to trust.

What to escalate.

That is the operating logic of the judgment economy.

And in that economy, clarity becomes one of the most valuable forms of work.

The Judgment Filter

Inputs pass through three gates: ignore, trust, escalate.

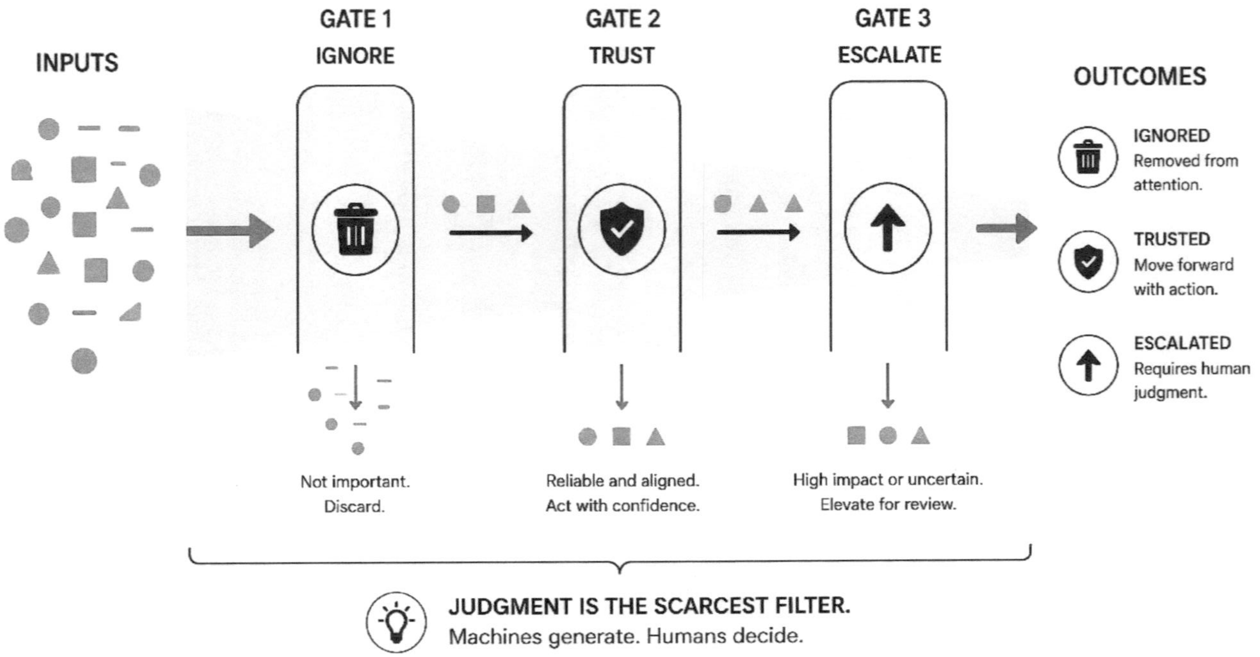

Chapter 16

The Taste Economy

For a long time, markets rewarded production.

The person who could manufacture more products often won. The person who could distribute more efficiently often won. The person who could produce more content, more services, more goods, more output, and more scale often gained an advantage.

This made sense in a world where production was difficult.

Scarcity creates value.

When something is difficult to make, the ability to make it becomes valuable.

The challenge of the coming era is that production itself is becoming increasingly abundant.

Artificial intelligence can generate content. Software can automate workflows. Robotics can automate labor. Digital systems can replicate experiences, products, recommendations, and services at extraordinary speed and scale.

The future may not suffer from a shortage of things.

It may suffer from a surplus.

A surplus of products.

A surplus of content.

A surplus of information.

A surplus of recommendations.

A surplus of opportunities.

A surplus of choices.

Abundance changes markets.

When production is scarce, the producer possesses leverage.

When production becomes abundant, selection possesses leverage.

This is the beginning of the taste economy.

The taste economy rewards those who can recognize what deserves to exist.

Not merely what can exist.

The distinction is important.

Technology increasingly expands what is possible.

Taste determines what is worthwhile.

A machine can generate a thousand product concepts. Taste determines which one belongs in the world.

A machine can generate a thousand advertisements. Taste determines which one strengthens the brand.

A machine can generate a thousand book covers. Taste determines which one communicates the right feeling.

A machine can generate a thousand business ideas. Taste determines which one deserves commitment.

The bottleneck shifts.

Creation becomes easier.

Recognition becomes harder.

For generations, many people assumed markets rewarded technical superiority. Sometimes they do.

But history suggests something more complicated.

Many technically superior products fail. Many technically inferior products succeed. Many brilliant ideas disappear. Many ordinary ideas thrive.

The difference often lies elsewhere.

Recognition.

Timing.

Positioning.

Human resonance.

Cultural fit.

Meaning.

These qualities sit closer to taste than engineering.

The market is not merely a machine for measuring capability.

It is a system for allocating attention.

And attention follows meaning.

This is why taste becomes economically significant.

The person with taste identifies significance before consensus arrives. They see value before value becomes obvious. They recognize quality before quality becomes popular. They identify alignment before the market fully understands what it is seeing.

This ability appears repeatedly throughout history.

A publisher chooses a manuscript that others rejected. An investor supports a company before the market understands it. A curator selects an artist before critics agree. A founder sees an unmet need before competitors notice it. A teacher notices potential before achievement appears.

In each case, taste functions as a form of recognition.

Not prediction.

Recognition.

The person sees something real before the crowd assigns value to it.

This becomes more important as artificial intelligence expands possibility.

Imagine a future where nearly every company can generate professional marketing. Every creator can generate images. Every entrepreneur can generate business plans. Every student can generate reports. Every consultant can generate recommendations.

The baseline quality rises.

The difference between average and excellent becomes harder to see.

This is where taste begins separating outcomes.

The future may not reward those who create the most.

It may reward those who create the most meaningful things.

The distinction sounds obvious.

It is not.

Modern systems often confuse quantity with significance. More content, more products, more features, more campaigns, more options, more growth, more visibility, and more output are often treated as indicators of progress.

Yet consumers rarely experience life this way.

Most people are not searching for more.

They are searching for better.

Better information.

Better experiences.

Better relationships.

Better solutions.

Better products.

Better stories.

Better ideas.

The challenge is that better requires judgment.

And judgment requires taste.

Taste becomes the mechanism through which quality emerges from abundance.

Without taste, abundance creates clutter.

With taste, abundance creates possibility.

This distinction changes the nature of competition.

The old economy rewarded efficiency.

The taste economy rewards discernment.

The old economy asked, *Can this be produced?*

The taste economy asks, *Should this be produced?*

The old economy rewarded execution.

The taste economy rewards selection.

The old economy valued capability.

The taste economy values significance.

This shift affects every field.

In business, founders increasingly face endless opportunities. Markets expand. Tools expand. Capabilities expand.

The challenge is no longer finding something to pursue.

The challenge is recognizing what belongs.

A founder with taste understands fit. Not every opportunity aligns with the business. Not every customer belongs. Not every expansion strengthens the company. Not every source of revenue improves the system.

Taste creates boundaries.

Those boundaries often produce better outcomes than unlimited growth.

The same principle applies to creativity.

Artificial intelligence can generate infinite variations.

But variation is not vision.

A creator with taste understands coherence. They know what belongs inside the work. They know what should remain absent. They know what strengthens the message. They know what dilutes it.

This ability becomes increasingly valuable because audiences are overwhelmed.

People do not merely want more content.

They want someone who can help them find what matters.

This is one reason trusted creators often gain influence.

They function as filters.

They save attention.

They reduce complexity.

They apply taste on behalf of others.

The same pattern appears in education.

The challenge is no longer access to information.

The challenge is determining which information deserves attention.

A student can learn almost anything.

The harder question is what is worth learning.

A teacher with taste helps answer that question. They distinguish foundational knowledge from distraction. They help students develop judgment about significance.

In a world of infinite options, this guidance becomes increasingly valuable.

The taste economy therefore extends far beyond art.

It shapes leadership.

Business.

Education.

Parenting.

Investing.

Design.

Innovation.

Relationships.

Even personal life.

Every day people make choices about what deserves attention, time, effort, and commitment.

Taste governs those choices.

This may be why taste is often misunderstood.

It appears subjective.

Yet its effects are objective.

Companies rise because of it.

Brands strengthen because of it.

Movements gain momentum because of it.

Careers advance because of it.

Cultures evolve because of it.

Taste helps determine which possibilities become realities.

That is power.

Not power through control.

Power through recognition.

The ability to recognize what deserves existence before the rest of the world agrees.

Artificial intelligence will likely make creation easier than at any point in human history.

This does not eliminate taste.

It increases its value.

The more possibilities that exist, the more important selection becomes. The more content that exists, the more important curation becomes. The more options that exist, the more important discernment becomes.

The more abundance expands, the more valuable recognition becomes.

This is why the future may not belong to the biggest producer.

It may belong to the person, company, teacher, creator, investor, leader, or parent who understands what belongs.

The one who can look at a thousand possibilities and recognize the handful worth pursuing.

The one who can separate novelty from significance.

The one who can distinguish attention from value.

The one who understands that not everything that can exist deserves to exist.

This is the operating principle of the taste economy.

Markets eventually reward those who see clearly.

Not merely those who produce endlessly.

Because abundance creates noise.

Taste creates meaning.

And meaning remains scarce.

Where scarcity remains, value follows.

The Taste Economy Funnel

Infinite options narrow into what deserves to exist.

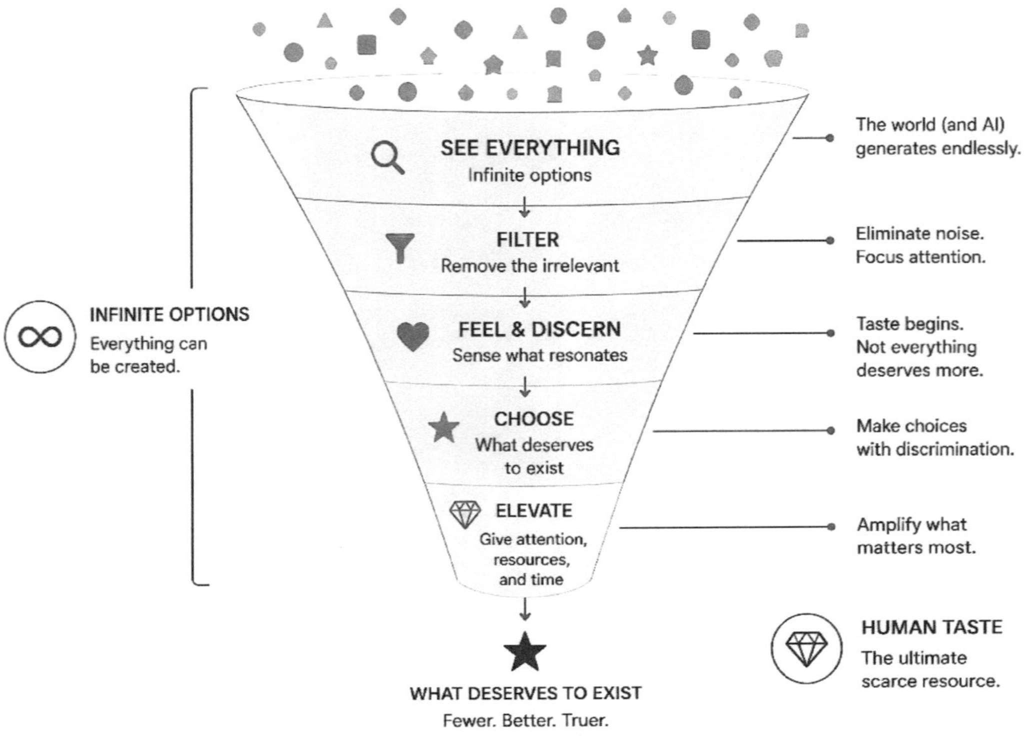

Chapter 17

The Trust Economy

Reputation used to be an advantage.

In the future, it may become infrastructure.

For a long time, reputation sat beside work rather than at its center. A person performed the service, built the product, gave the advice, delivered the result, and over time other people formed an opinion.

Reliable.

Difficult.

Brilliant.

Careless.

Honest.

Expensive.

Generous.

Unstable.

Reputation followed behavior.

It traveled slowly through word of mouth, references, introductions, repeat business, and memory. It was personal, local, and relational.

A person's reputation mattered, but it rarely appeared on the balance sheet. It did not fit neatly into a contract. It did not show up in a dashboard. It was difficult to measure and often invisible to formal systems.

Yet it quietly shaped everything.

Who received the call.

Who got invited into the room.

Who was believed.

Who received the benefit of the doubt.

Who was trusted with risk.

Who was allowed to move faster because others did not feel the need to verify every step.

That is the hidden power of trust.

Trust reduces friction.

When trust is high, fewer explanations are required. Fewer safeguards are needed. Fewer meetings become necessary. Fewer doubts interrupt the work.

People move.

They decide.

They cooperate.

They extend grace.

They share information.

They bring opportunities forward.

A trusted person does not merely possess a good reputation.

They become part of the operating system.

This is what changes in the trust economy.

As information becomes abundant, trust becomes more valuable. As synthetic content expands, trust becomes more necessary. As artificial intelligence generates more answers, more documents, more images, more strategies, more recommendations, and more claims, people will increasingly ask a simpler question:

Who do I trust?

That question becomes infrastructure.

It determines what enters attention. It determines what receives belief. It determines what moves forward. It determines what gets funded, purchased, read, shared, hired, followed, protected, and forgiven.

In a world where anyone can generate the appearance of expertise, reputation becomes a filter.

Not a decorative layer.

A structural one.

The trusted person saves others the cost of constant verification. The trusted company saves clients the cost of constant supervision. The trusted creator saves audiences the cost of sorting through endless noise. The trusted leader saves teams the cost of second-guessing every decision.

Trust turns uncertainty into movement.

This is why reputation compounds.

One trustworthy act matters.

But repeated trustworthy acts create infrastructure.

A person answers honestly once. Then again. Then under pressure. Then when it costs them something. Then when no one is watching.

Over time, others begin to rely on the pattern.

The pattern becomes reputation.

The reputation becomes trust.

The trust becomes access.

The access becomes opportunity.

The opportunity becomes leverage.

This is not manipulation.

It is the natural compounding of integrity.

The trust economy rewards people whose behavior can bear repetition.

Not perfection.

Consistency.

People do not need leaders, advisors, creators, friends, or institutions to be flawless. They need them to be real, responsible, and steady enough that trust can survive complexity.

This matters because the future will create more complexity than most people can evaluate directly.

No one will be able to personally verify every claim, inspect every system, test every tool, research every source, or understand every recommendation. The volume will simply be too great.

Trust becomes the bridge across the impossible scale of modern life.

But trust is fragile.

It is built slowly and lost quickly.

This is why reputation becomes infrastructure rather than branding.

Branding can be designed.

Reputation must be earned.

Branding can create attention.

Reputation sustains belief.

Branding can open a door.

Reputation determines whether people stay.

The difference becomes increasingly important in a synthetic world.

Artificial intelligence can help create a polished brand almost instantly. It can generate language, visuals, campaigns, slogans, posts, websites, proposals, presentations, and professional identity.

The surface can appear credible before the substance exists.

This creates a market problem.

When appearances become easier to manufacture, people begin valuing evidence of continuity.

Who has shown up over time?

Who has told the truth when truth was inconvenient?

Who has delivered when no one was applauding?

Who has protected the standard when shortcuts were available?

Who has acted with alignment when incentives encouraged distortion?

These questions belong to the trust economy.

Reputation becomes the record of how someone behaves when performance alone is not enough.

It is not merely what a person says.

It is what repeated behavior teaches others to expect.

This applies to individuals, companies, institutions, books, brands, schools, families, platforms, leaders, and communities.

A company with strong trust can move through uncertainty because clients believe it will handle what arises. A leader with strong trust can make difficult decisions because people believe the decision is not merely self-serving. A creator with strong trust can introduce new ideas because readers believe the work emerges from a coherent body of thought.

A parent with strong trust can guide a child because the child believes correction is connected to care.

Trust changes how information lands.

The same sentence spoken by an untrusted person may sound like manipulation.

Spoken by a trusted person, it may sound like guidance.

The content did not change.

The context did.

That context is reputation.

In the automated age, reputation also becomes protection against commoditization.

If machines can produce acceptable outputs, output alone becomes easier to compare, copy, and replace.

Trust is harder to copy.

A competitor may copy a service.

They cannot instantly copy a decade of reliability.

A machine may imitate a writing style.

It cannot inherit the lived relationship between an author and a reader.

A platform may recommend alternatives.

It cannot automatically transfer trust.

This is why the trust economy rewards accumulated credibility.

It rewards the long pattern.

The record.

The visible history of alignment between words and behavior.

In a world of shortcuts, the slow-built asset becomes more valuable.

That asset is trust.

The mistake is assuming trust is soft.

Trust is operational.

It determines speed.

It determines cost.

It determines access.

It determines resilience.

It determines whether people cooperate under uncertainty.

Low-trust systems are expensive.

They require more contracts, more oversight, more monitoring, more meetings, more approvals, more proof, and more control.

High-trust systems move differently.

They do not eliminate structure, but they reduce drag.

People share information sooner. They surface problems earlier. They assume good faith. They recover from mistakes faster. They spend less energy protecting themselves from one another.

Trust is therefore not separate from performance.

Trust is part of performance.

It is the invisible architecture that allows work to move.

This is why reputation becomes infrastructure.

It is not an afterthought.

It is a load-bearing system.

The future will likely produce more generated content, more automated recommendations, more synthetic identities, and more scalable appearances of competence.

That will not make trust obsolete.

It will make trust more valuable.

People will increasingly seek trusted filters.

Trusted advisors.

Trusted creators.

Trusted brands.

Trusted communities.

Trusted leaders.

Trusted institutions.

Not because they want less intelligence.

Because they need a way to navigate too much of it.

The trust economy begins when people stop asking only, *What does this say?* and begin asking, *Who stands behind it?*

Who has earned belief?

Who has carried responsibility?

Who has behaved consistently enough to reduce uncertainty?

Who can be relied upon when the situation becomes complicated?

Those questions cannot be answered instantly.

They require time.

They require pattern.

They require memory.

They require reputation.

And in an age where so much can be generated on demand, the things that cannot be generated on demand become increasingly valuable.

Trust is one of them.

Reputation is the visible trace of trust earned over time.

That trace becomes infrastructure.

It holds relationships.

It holds markets.

It holds institutions.

It holds communities.

It holds the fragile bridge between abundance and belief.

In the trust economy, the most valuable people are not merely those who can produce.

They are those others can rely on when production is no longer rare.

That reliability becomes the asset.

And once earned, it becomes one of the strongest forms of human advantage.

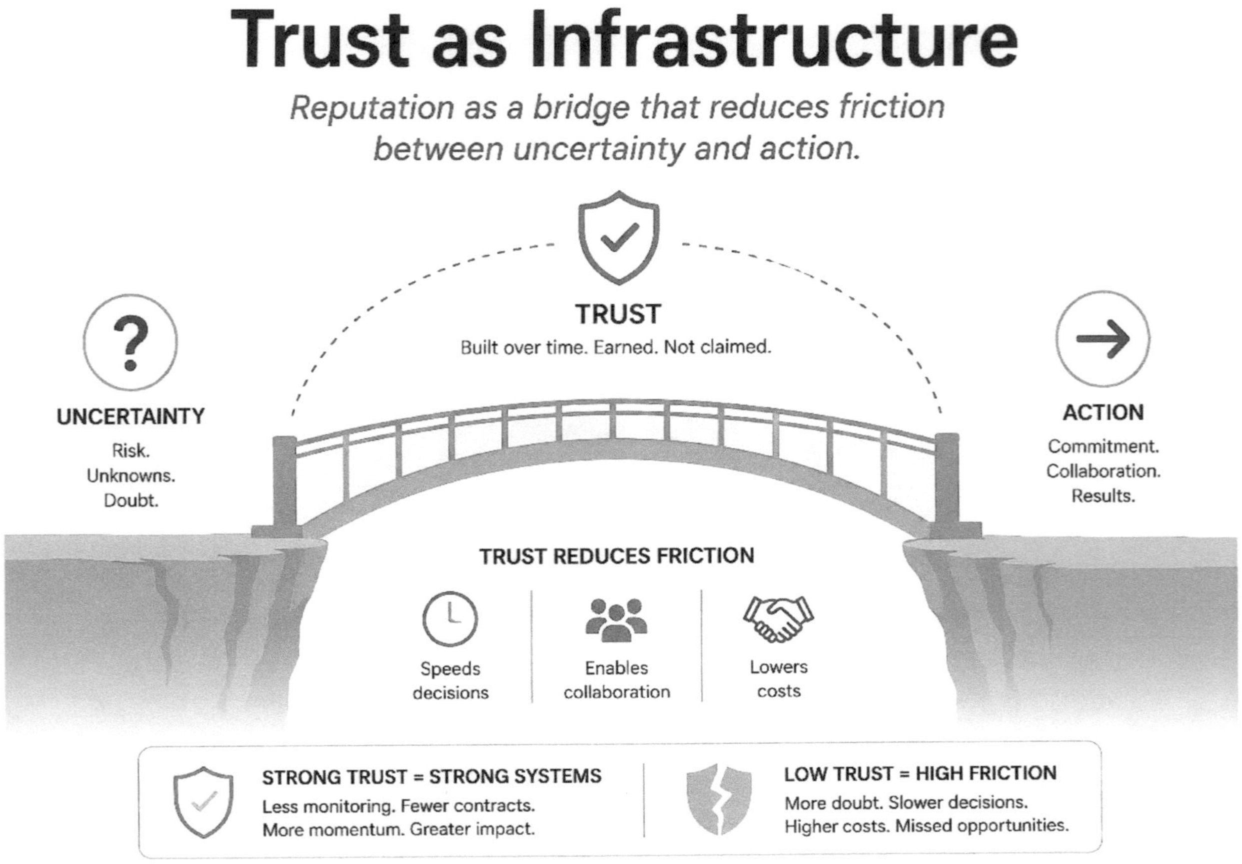
Trust as Infrastructure
Reputation as a bridge that reduces friction between uncertainty and action.
TRUST
Built over time. Earned. Not claimed.
UNCERTAINTY
Risk.
Unknowns.
Doubt.
ACTION
Commitment.
Collaboration.
Results.
TRUST REDUCES FRICTION
Speeds decisions
Enables collaboration
Lowers costs
STRONG TRUST = STRONG SYSTEMS
Less monitoring. Fewer contracts.
More momentum. Greater impact.
LOW TRUST = HIGH FRICTION
More doubt. Slower decisions.
Higher costs. Missed opportunities.

Part IV

Life After Efficiency

Chapter 18

Efficiency Is Not the Highest Human Good

Every age develops an ideal.

For much of the modern era, that ideal has been efficiency.

Faster.

Cheaper.

Smaller.

More scalable.

More productive.

More optimized.

More streamlined.

More automated.

More measurable.

More output from fewer inputs.

The logic is compelling.

Efficiency reduces waste. Efficiency reduces cost. Efficiency frees time. Efficiency expands capability. Efficiency allows societies to produce more food, build more homes, move more people, distribute more information, and solve more problems.

Many of the comforts of modern life exist because human beings became extraordinarily skilled at making systems more efficient.

The problem is not efficiency.

The problem begins when efficiency stops being a tool and becomes a philosophy.

When efficiency becomes the highest good, something important starts to disappear.

Not everything valuable can be measured by output. Not everything meaningful should be accelerated. Not everything human benefits from optimization.

This distinction may become increasingly important as artificial intelligence and automation expand.

Machines excel at efficiency. They process faster, sort faster, calculate faster, generate faster, respond faster, and scale faster.

The future will almost certainly become more efficient.

The question is whether human beings will remember that efficiency and value are not always the same thing.

Consider a conversation.

An efficient conversation would deliver information quickly. Questions would be answered immediately. Decisions would be reached rapidly. No unnecessary words would be exchanged. No detours would occur. No silence would exist.

Yet most people would recognize that many of the most meaningful conversations in life do not work this way.

They wander.

They pause.

They repeat.

They reflect.

They circle back.

They reveal themselves slowly.

The value is not only in the information exchanged.

The value is in the relationship formed.

Efficiency is not the highest good.

The same pattern appears throughout life.

An efficient meal delivers nutrition.

A meaningful meal creates connection.

An efficient book transfers information.

A meaningful book changes perception.

An efficient walk reaches a destination.

A meaningful walk creates observation.

An efficient childhood prepares for adulthood.

A meaningful childhood includes wonder.

An efficient friendship solves problems.

A meaningful friendship shares life.

The distinction matters because modern systems often optimize what is easiest to measure.

Time.

Cost.

Speed.

Output.

Volume.

Productivity.

These metrics are useful.

But they are incomplete.

Many of the most important human experiences resist measurement.

Trust.

Belonging.

Love.

Meaning.

Creativity.

Presence.

Wonder.

Grief.

Joy.

These realities often unfold according to a different logic.

Not the logic of efficiency.

The logic of significance.

Significance asks different questions.

Not *How quickly can this happen?*

But *What is happening?*

Not *How much can we produce?*

But *What are we producing?*

Not *How do we reduce the time?*

But *What deserves the time?*

This difference becomes increasingly important in an age of powerful technology.

Artificial intelligence may allow people to produce more content than ever before.

The question becomes whether more content creates more meaning.

Automation may reduce the effort required for many tasks.

The question becomes what people do with the energy that returns.

Robotics may eliminate friction throughout daily life.

The question becomes whether a frictionless life is automatically a better life.

History suggests caution.

Many human experiences derive value from effort.

Not all effort.

Meaningful effort.

The effort of learning a skill.

The effort of building trust.

The effort of creating something original.

The effort of caring for another person.

The effort of recovering after failure.

The effort of remaining loyal when loyalty is inconvenient.

These experiences shape character precisely because they cannot be optimized away completely.

A world obsessed with efficiency sometimes forgets this.

It assumes that difficulty is always a problem.

Sometimes difficulty is part of the value.

A mountain climbed feels different from a mountain viewed. A language learned feels different from a translation received. A friendship maintained feels different from a connection suggested by an algorithm. A garden grown feels different from a garden purchased.

The effort becomes part of the meaning.

This does not mean all effort is valuable.

Many forms of labor are repetitive, wasteful, exhausting, and unnecessary. Technology should absolutely reduce pointless friction.

The goal is not to romanticize inefficiency.

The goal is to distinguish between burdens that diminish life and experiences that deepen it.

The future may require greater wisdom in making that distinction.

For decades, many organizations treated efficiency as the primary measure of success. The most efficient company won. The most efficient process won. The most efficient workflow won.

Yet efficiency often creates hidden tradeoffs.

A customer service system may become more efficient while feeling less human. A school may become more efficient while becoming less educational. A hospital may become more efficient while becoming less healing. A workplace may become more efficient while becoming less trusting.

The optimization succeeded.

Something else weakened.

This is the danger of elevating efficiency above purpose.

The system becomes increasingly good at doing something without asking whether the thing itself still deserves to be done.

Efficiency is a multiplier.

It amplifies direction.

If the direction is wise, efficiency helps.

If the direction is unwise, efficiency accelerates the mistake.

This is why meaning must govern efficiency rather than the reverse.

The future may tempt people toward continuous optimization.

Every activity can be measured. Every habit can be tracked. Every workflow can be improved. Every decision can be informed by data. Every moment can be used. Every process can be accelerated.

The temptation is subtle.

A person begins optimizing work.

Then sleep.

Then health.

Then relationships.

Then hobbies.

Then leisure.

Then creativity.

Eventually life becomes a project.

A system.

A dashboard.

A performance report.

And something essential begins to disappear.

The space for wonder.

The space for spontaneity.

The space for observation.

The space for play.

The space for stillness.

The space for experiences that are valuable precisely because they serve no measurable purpose.

Human beings do not thrive only through achievement.

They thrive through meaning.

Meaning often requires time that cannot be justified through efficiency alone.

A parent reading a story to a child is not maximizing productivity. A friend sitting beside another friend during grief is not maximizing productivity. A person watching a sunset is not maximizing productivity. A musician practicing simply for the joy of it is not maximizing productivity. A writer wandering through an idea before it becomes a book is not maximizing productivity.

Yet many of the richest experiences in life emerge from exactly these moments.

Efficiency struggles to explain them.

Meaning does not.

This may become one of the defining questions of the post-automation world.

If machines can increasingly handle the tasks of efficiency, what remains for human beings?

The answer is not endless consumption.

It is not infinite productivity.

It is not optimizing every corner of existence.

It may be learning how to inhabit life more fully.

How to pay attention.

How to create.

How to love.

How to belong.

How to care.

How to reflect.

How to experience reality directly rather than merely process it.

Technology can help create the conditions for this.

But technology cannot decide it.

The decision remains human.

A machine can free an hour.

Only a human can determine what that hour is for.

That question becomes increasingly important as efficiency expands.

Because every increase in capability creates a corresponding increase in freedom.

And freedom always asks a deeper question.

Now that you can do more—

What deserves doing?

This is where efficiency reaches its limit.

It can tell us how.

It cannot tell us why.

Meaning answers the question efficiency cannot.

And in the end, the quality of a human life may depend less on how efficiently it was lived than on whether it was lived in service of what truly mattered.

Efficiency is valuable.

But it is not the highest human good.

The future will belong to those who remember the difference.

Chapter 19

The Return of the Real

Every technological revolution creates a counter-movement.

The more industrialized the world became, the more people valued handmade objects. The more communication became digital, the more people valued face-to-face conversation. The more music became available everywhere, the more people valued live performance. The more photography became effortless, the more people valued original art.

The pattern repeats because abundance changes perception.

When something becomes common, attention shifts toward what remains scarce.

This is one of the central paradoxes of the coming age.

As artificial intelligence and automation expand, many people assume the future will become increasingly virtual, increasingly synthetic, increasingly digital, and increasingly detached from physical reality.

That may be partially true.

But another trend may emerge at the same time.

The return of the real.

Not because technology fails.

Because technology succeeds.

The more convincing synthetic experiences become, the more valuable authentic experiences become.

The more artificial intelligence can generate books, images, videos, conversations, music, recommendations, and simulated environments, the more people may find themselves longing for something else.

Something slower.

Something embodied.

Something earned.

Something that exists independently of the screen.

The return of the real is not a rejection of technology.

It is a rebalancing.

A reminder that human beings do not live entirely in the world of information.

They live in the world of experience.

And experience possesses qualities that cannot be fully replicated.

This is why certain things may gain new significance.

Books.

Gardens.

Meals.

Music.

Conversation.

Craft.

Place.

At first glance, these seem unrelated.

Yet they share something important.

They anchor people in reality.

A book is not merely information.

It is attention embodied.

Someone sat with an idea long enough to shape it. A reader sits with it long enough to receive it. The exchange unfolds slowly and asks something from both people: patience, reflection, and presence.

These qualities may become increasingly rare.

And therefore increasingly valuable.

The same is true of a garden.

A garden refuses acceleration.

It grows according to biological time rather than market time, algorithmic time, or quarterly-report time. Seeds do not respond to urgency. Roots do not care about productivity metrics.

Growth follows a different rhythm.

The gardener participates in that rhythm.

The experience teaches patience because reality itself imposes patience.

No amount of technology can fully remove that lesson.

Meals function similarly.

An efficient meal delivers calories.

A meaningful meal creates relationship.

People gather. Stories emerge. Silence appears. Laughter appears. Memory forms.

The food matters.

But the experience often matters more.

This is why family dinners, community meals, and shared tables continue to hold significance even when food delivery becomes increasingly convenient.

The value is not only nourishment.

The value is presence.

Music offers another example.

The future may contain infinite generated songs, infinite playlists, infinite personalized soundtracks, and infinite musical variations.

Yet live music may become more valuable rather than less.

Not because it is technically superior.

Because it is real.

A musician stands in front of an audience. Mistakes are possible. Variation is inevitable. The moment exists only once.

People gather to experience something together that cannot be perfectly reproduced.

Reality creates meaning through irreversibility.

The moment matters because it cannot be replayed exactly.

Conversation may become even more important.

Artificial intelligence can already answer questions, generate responses, provide companionship, and sustain dialogue. These capabilities will improve dramatically.

Yet genuine human conversation contains something different.

Not merely information exchange.

Mutual vulnerability.

Shared memory.

Unexpected discovery.

A sense that another consciousness is present.

Not simulated.

Present.

Human beings have always needed this.

The need does not disappear when technology advances.

It may become more visible.

Craft follows the same pattern.

The industrial age rewarded mass production. The digital age rewarded scale. The automated age may create renewed appreciation for human craftsmanship.

Not because handcrafted objects are always more efficient.

They are usually less efficient.

But efficiency is not the only source of value.

Craft carries evidence of attention, care, effort, and judgment.

The object contains traces of the person who made it.

That trace matters.

It connects people across time.

A handcrafted table.

A handmade ceramic bowl.

A stitched quilt.

A handwritten note.

A carefully bound book.

These objects remind people that not everything valuable emerges from automation.

Some value emerges from human presence itself.

Place may become one of the most significant returns of all.

Technology increasingly allows people to exist anywhere while physically remaining nowhere. Work can happen remotely. Relationships can happen digitally. Entertainment can happen privately. Information can arrive instantly.

The advantages are substantial.

Yet place still matters.

A neighborhood.

A library.

A coffee shop.

A local bookstore.

A park.

A church.

A school.

A creek.

A city square.

A home.

These locations provide something digital systems struggle to reproduce.

Context.

History.

Belonging.

Identity.

People are shaped by places.

Places carry memory.

The future may therefore produce renewed appreciation for local experience.

Not because global connection disappears.

Because local reality remains important.

A person can participate in a global network while still needing a place where they are known.

The return of the real is ultimately about friction.

Not harmful friction.

Meaningful friction.

The friction of growing something.

Learning something.

Building something.

Practicing something.

Maintaining something.

Caring for someone.

Showing up repeatedly.

Modern systems often seek to remove friction.

Many forms of friction should be removed.

But some forms of friction create depth.

A friendship deepens through time.

Trust deepens through repetition.

Mastery deepens through practice.

Identity deepens through lived experience.

Automation cannot eliminate these realities without changing their nature.

This may explain why people increasingly seek experiences that feel tangible.

Physical books.

Farmer's markets.

Gardens.

Craft fairs.

Local restaurants.

Independent bookstores.

Vinyl records.

Handmade goods.

Community events.

These trends are not nostalgia.

They are adaptation.

People intuitively seek what remains scarce.

Reality is becoming scarce in certain forms.

Not absent.

Scarce.

The return of the real is therefore not anti-technology.

It is pro-human.

It recognizes that technology works best when it supports life rather than replaces it.

A machine can order the groceries.

The family still shares the meal.

A machine can generate the itinerary.

The traveler still walks the street.

A machine can summarize the book.

The reader still experiences the insight.

A machine can compose the music.

The musician still performs.

A machine can optimize the schedule.

The friend still shows up.

The future may contain extraordinary virtual worlds, extraordinary synthetic experiences, extraordinary automation, and extraordinary intelligence.

Yet some of the most valuable experiences may look surprisingly familiar.

A book read in a quiet chair.

A garden growing through the seasons.

A meal shared with people who matter.

Music filling a room.

A conversation that changes someone.

A craft practiced carefully.

A place that feels like home.

These things endure because they satisfy needs deeper than efficiency.

They connect people to reality.

To one another.

To time.

To memory.

To meaning.

The return of the real is not a step backward.

It may be one of the most important steps forward.

Because as artificial experiences multiply, authentic experiences become more precious.

And wherever something precious becomes scarce, human beings eventually learn to value it again.

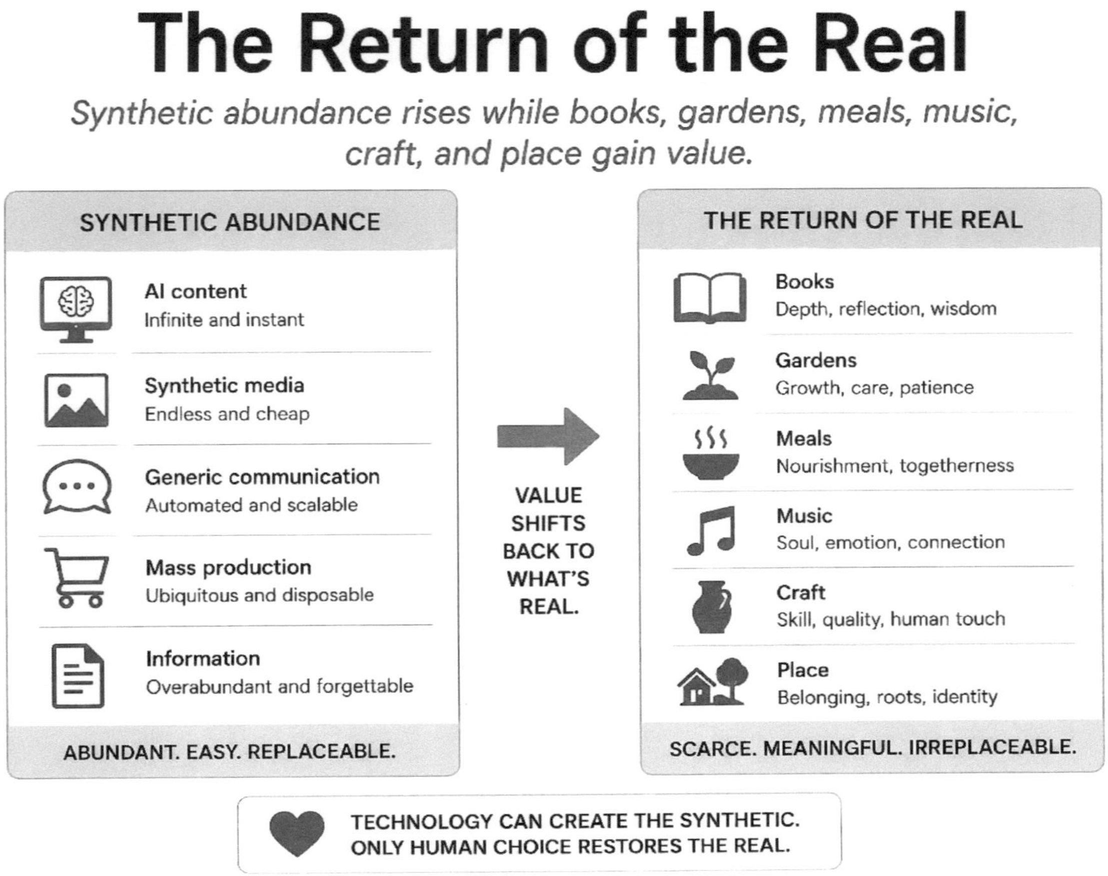
The Return of the Real
Synthetic abundance rises while books, gardens, meals, music, craft, and place gain value.
SYNTHETIC ABUNDANCE
AI content
Infinite and instant
Synthetic media
Endless and cheap
Generic communication
Automated and scalable
Mass production
Ubiquitous and disposable
Information
Overabundant and forgettable
ABUNDANT. EASY. REPLACEABLE.
VALUE SHIFTS BACK TO WHAT'S REAL.
THE RETURN OF THE REAL
Books
Depth, reflection, wisdom
Gardens
Growth, care, patience
Meals
Nourishment, togetherness
Music
Soul, emotion, connection
Craft
Skill, quality, human touch
Place
Belonging, roots, identity
SCARCE. MEANINGFUL. IRREPLACEABLE.
TECHNOLOGY CAN CREATE THE SYNTHETIC.
ONLY HUMAN CHOICE RESTORES THE REAL.

Chapter 20

Human Energy as a Sacred Resource

Automation returns something to human beings.

At least, it should.

It should return time. It should return attention. It should return physical capacity. It should return mental space. It should return the parts of life that repetitive labor once consumed.

This is the promise beneath the tools.

The promise is not merely that more can be produced.

The promise is that less human energy must be spent on work that does not require a human soul.

Yet modern systems often misunderstand this.

They treat every recovered minute as capacity for additional output. If a machine saves an hour, the system immediately asks what else can be done in that hour. If software reduces effort, volume increases. If automation removes friction, expectations rise.

The human being does not receive the recovered life.

The machine creates efficiency, and the system converts the efficiency into more demand.

This is not liberation.

It is acceleration.

The purpose of automation should not be endless output.

It should be recovered life.

This distinction matters because human energy is not infinite.

A person can only think deeply for so long. Care for others for so long. Decide under pressure for so long. Create from depth for so long. Respond to complexity for so long.

Human beings are not machines with softer surfaces.

They are living systems.

Living systems require rest, silence, recovery, unstructured time, emotional renewal, physical restoration, meaning, and relationship.

Without these things, the system may continue producing, but the person begins disappearing inside the production.

Many modern organizations have learned how to measure output while ignoring depletion.

The work gets done.

The messages get answered.

The reports get generated.

The meetings occur.

The metrics improve.

Yet the human operating system underneath begins to weaken.

Attention fragments.

Judgment declines.

Irritability rises.

Creativity narrows.

Trust thins.

The system appears productive until it suddenly becomes brittle.

This is what happens when human energy is treated as an endlessly renewable input.

It is not.

Human energy is a sacred resource.

Sacred does not mean fragile in a decorative sense.

It means worthy of protection because it carries life.

Human energy is the source of judgment, care, creativity, courage, presence, and responsibility. It is the energy a parent uses to notice a child. The energy a leader uses to make a difficult decision. The energy a creator uses to bring something original into the world. The energy a friend uses to remain present during suffering.

It is the energy a person uses to become more than a function.

If that energy is exhausted, the human advantage weakens.

This is why automation must be governed by a human question.

What should the saved energy serve?

If the answer is always more output, automation becomes another form of extraction.

The tools become more advanced.

The life does not become more whole.

The person moves faster. The calendar fills again. The expectations rise again. The nervous system remains flooded. The recovered time disappears into the next demand.

This is one of the great risks of the automated age.

Technology may reduce tasks without reducing exhaustion. It may increase capacity without increasing freedom. It may expand production without expanding life.

The difference lies in the operating system surrounding the tool.

A task removed from human life creates a possibility.

Not a guarantee.

The possibility may become rest.

Or creation.

Or relationship.

Or learning.

Or contemplation.

Or play.

Or health.

Or presence.

Or it may simply become more work.

The choice matters.

A culture that worships productivity will use automation to intensify productivity. A culture that values human flourishing will use automation to protect life.

This choice appears everywhere.

In companies.

In families.

In schools.

In personal routines.

In public policy.

In private ambition.

The question remains the same.

When technology gives energy back, who receives it?

The system?

Or the human being?

This question becomes especially important for people who are naturally driven.

The founder.

The creator.

The parent.

The professional.

The person whose mind never stops building.

Automation can become a multiplier of purpose.

It can also become a multiplier of depletion.

A person with powerful tools may produce at an extraordinary pace and mistake that pace for health. They may confuse capacity with sustainability. They may assume that because output is possible, output is required.

But human energy has rhythms.

Even the most creative mind needs fallow ground. Even the strongest body needs recovery. Even the most capable leader needs silence. Even the most loving parent needs restoration.

This is not weakness.

It is design.

Living systems renew through cycles.

Exertion.

Recovery.

Expansion.

Rest.

Creation.

Integration.

The machine may operate continuously.

The human being cannot.

When humans imitate machines, they become less human.

The goal of automation should not be to make people operate without pause.

The goal should be to free people from unnecessary depletion so they can offer their best energy where it matters most.

This requires discernment.

Not every task deserves equal human attention.

Some tasks can be automated.

Some can be delegated.

Some can be simplified.

Some can be eliminated.

Some must remain human.

The work of the future is not merely doing more.

It is deciding where human energy belongs.

Human energy belongs where meaning is created. Where judgment is required. Where trust is formed. Where relationships are sustained. Where responsibility is carried. Where beauty is recognized. Where courage is needed. Where presence changes the outcome.

Machines can support these spaces.

They should not consume the energy meant for them.

This is why rest becomes strategic.

Not rest as collapse.

Rest as stewardship.

A rested human sees more clearly. A rested human chooses more wisely. A rested human creates from depth rather than urgency. A rested human listens better. A rested human notices what a depleted human misses.

In an age of automation, rest may become one of the most misunderstood forms of advantage.

The exhausted person uses tools reactively.

The rested person uses tools intentionally.

The exhausted person automates to keep up.

The rested person automates to choose better.

The exhausted person fills every recovered minute.

The rested person protects some recovered minutes as life.

This difference compounds.

A life built around endless output eventually narrows.

A life built around recovered energy can deepen.

The question is not whether technology can make us more productive.

It can.

The question is whether productivity remains in service of life.

If automation frees a person from repetitive effort but steals the possibility of rest through higher expectations, then the system has not restored human life.

It has merely upgraded the machinery of demand.

The Human Advantage Operating System rejects that arrangement.

It treats human energy as something to be governed, not consumed. It asks whether tools are returning people to what matters or pulling them further away. It asks whether efficiency is creating freedom or merely raising the speed limit. It asks whether recovered time is being converted into presence, creativity, care, and meaning.

This may be one of the central moral questions of the automated age.

What will humanity do with the energy automation returns?

Will we produce endlessly?

Or will we recover life?

The answer will determine whether technology makes human beings more replaceable or more fully human.

Because the purpose of automation is not to remove the human from life.

It is to remove unnecessary burden so the human can return to life.

That distinction must be protected.

Human energy is not merely a resource for production.

It is the source of everything production is supposed to serve.

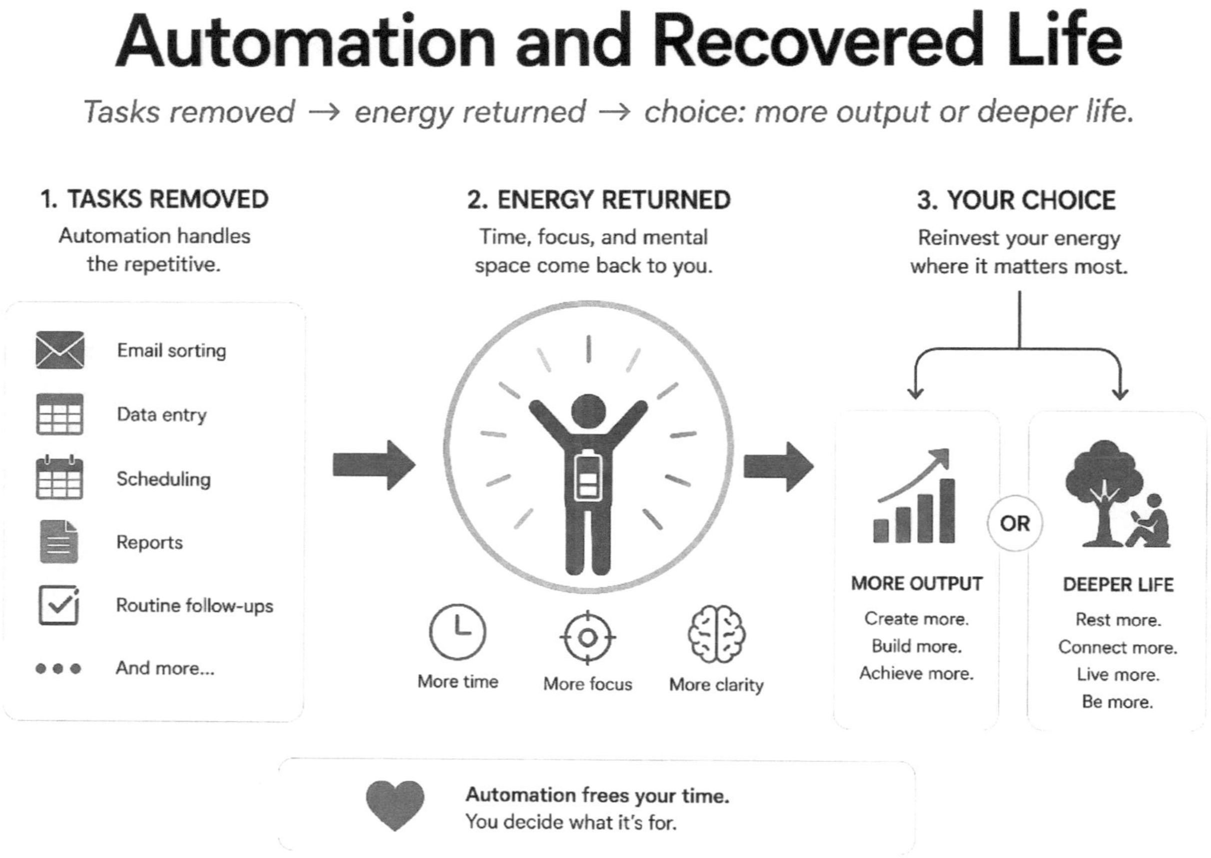
Automation and Recovered Life
Tasks removed → energy returned → choice: more output or deeper life.
1. TASKS REMOVED
Automation handles the repetitive.
Email sorting
Data entry
Scheduling
Reports
Routine follow-ups
And more...
2. ENERGY RETURNED
Time, focus, and mental space come back to you.
More time
More focus
More clarity
3. YOUR CHOICE
Reinvest your energy where it matters most.
MORE OUTPUT
Create more.
Build more.
Achieve more.
OR
DEEPER LIFE
Rest more.
Connect more.
Live more.
Be more.
Automation frees your time.
You decide what it's for.

Chapter 21

The Creative Human

For centuries, creativity was constrained by production.

An artist could paint only so many paintings. A writer could write only so many books. A musician could compose only so many songs. Every act of creation required time, effort, skill, and attention. The challenge was making the thing.

Artificial intelligence changes the location of that challenge.

The bottleneck is no longer production.

The bottleneck is recognition.

A machine can generate images, stories, music, designs, and ideas at a scale no human being could ever match. It can create possibilities endlessly. For some people, this creates anxiety. If machines can create, what remains for human creators?

The deeper question is not whether machines can create.

The deeper question is what kind of creation remains uniquely human.

To answer that question, it helps to distinguish between production and origination.

Production brings something into existence.

Origination brings something new into existence.

A machine can generate variations. A human being can notice something that has not yet been named. A machine can recombine patterns. A human being can experience a realization. A machine can imitate style. A human being can develop a point of view.

The future creator may become less valuable as a producer and more valuable as an originator.

This is not a reduction of creativity.

It is a clarification of where creativity begins.

For many years, people assumed creativity meant making things. Writing the book. Painting the painting. Composing the music. Building the product. Designing the campaign.

Those activities still matter.

But beneath them lies something deeper.

Observation.

Most creative work begins with noticing.

Someone notices a pattern that others overlook. Someone notices a contradiction, a possibility, a tension, or a truth hiding in plain sight. The observation comes first. The creation follows.

A machine can generate thousands of answers.

It cannot independently decide which question feels strangely important.

A machine can produce endless content.

It cannot experience the moment when a person suddenly sees the world differently.

This matters because many of the most important creative breakthroughs begin as observations rather than outputs. A scientist notices an anomaly. A writer notices a pattern. An entrepreneur notices a gap. An artist notices a feeling. A teacher notices a need.

The act of noticing creates the possibility of something new.

This is one reason creativity may become more valuable rather than less valuable as artificial intelligence expands.

When production becomes abundant, originality becomes scarce.

And scarcity creates value.

The future may contain an endless supply of content. What it may lack is genuinely original perception.

A machine can tell a story.

But who notices the story worth telling?

A machine can create a product.

But who recognizes the unmet human need?

A machine can write a book.

But who sees the question that deserves exploration?

Origination remains deeply connected to human experience.

Not because humans are more efficient creators.

Because human beings live inside realities that machines do not.

Wonder.

Curiosity.

Love.

Grief.

Memory.

Failure.

Hope.

Regret.

Belonging.

These experiences create perspective.

Perspective creates originality.

A machine may understand descriptions of grief. A human being experiences grief. A machine may describe courage. A person chooses courage despite fear. A machine may explain love. A human being lives it.

Information contributes.

Experience transforms.

This is why the future creator becomes less valuable for the volume of output and more valuable for the depth of perception.

Modern systems often reward production. Publish more. Post more. Create more. Launch more. Share more.

That logic made sense when production was scarce.

But abundance changes the equation.

When everyone can produce, production becomes less differentiating.

What remains valuable is the ability to see.

The creator becomes a witness.

A translator of reality.

Someone who notices what others miss and gives it form.

This may explain why certain books, songs, paintings, and ideas endure. Not because they were technically difficult to produce, but because they revealed something true.

Truth creates longevity.

Novelty creates attention.

The two are not the same.

Artificial intelligence will likely generate unprecedented amounts of novelty. The human creator may increasingly become responsible for truth—not factual truth alone, but experiential, psychological, relational, moral, and existential truth.

The truths people recognize because they have lived them.

This is where creativity becomes connected to meaning.

The creator is not simply generating objects.

The creator is helping people see.

A good book changes perception. A good piece of music changes perception. A good conversation changes perception. A good teacher changes perception. The deepest creative acts reorganize how people understand reality.

Machines may help with production.

Origination still begins somewhere human.

This shift matters far beyond art.

Organizations need creative humans. Communities need creative humans. Schools need creative humans. Families need creative humans.

The future will continuously generate new situations, new technologies, new challenges, and new questions. Old answers will not always fit. Someone must originate new ones.

That responsibility belongs to people willing to perceive differently.

The future therefore rewards a specific form of creativity.

Not creativity as endless output.

Creativity as meaningful observation.

The ability to stand in the world, notice something real, and bring forth a perspective that did not previously exist.

This is increasingly difficult because ideas are not scarce.

Attention is.

The creator must protect curiosity, wonder, and the ability to notice.

Many creative breakthroughs emerge during activities that appear unproductive—walking, reading, reflecting, observing, listening, exploring, and resting. These moments often look inefficient from the outside.

Yet they are where origination frequently begins.

A machine can produce continuously.

A human creator often requires stillness.

This may create a surprising advantage for those who remain capable of sustained observation. The person who notices deeply may outperform the person who produces endlessly.

Because the source of creativity is not production.

The source is perception.

This is why the creative human remains essential.

Not because machines are incapable.

Because machines do not independently care.

They do not wonder. They do not long. They do not experience beauty, loss, love, aging, parenthood, mortality, or the strange mixture of joy and sorrow that often gives rise to the most meaningful forms of creativity.

Human beings do.

And from those experiences emerge questions, stories, ideas, products, movements, and works that did not previously exist.

The future will likely contain more production than any civilization has ever known.

What it may desperately need are people who can originate.

People who can notice.

People who can see.

People who can imagine.

People who can bring forth something meaningful from the raw material of human experience.

Machines can produce.

Humans must originate.

That is not a limitation.

It may be one of the deepest forms of human advantage.

Chapter 22

The Relational Human

The future will still require people who know how to stay.

This may sound simple.

It is not.

Many modern systems are designed around movement. Faster response. Faster delivery. Faster creation. Faster decisions. Faster replacement. Faster exit when something becomes inconvenient.

Technology increases speed. Markets reward speed. Platforms normalize speed. People become accustomed to speed.

But relationship does not mature at the speed of automation.

Care takes time. Presence takes attention. Forgiveness takes humility. Loyalty takes endurance.

These are not minor human qualities left over from an older world.

They are part of what makes human life livable.

A machine can respond.

A person can care.

The difference is not always visible at first. A response may sound caring. A generated message may be warm. A system may remember a preference, offer comfort, provide encouragement, and simulate concern.

These things may be useful.

But care is not merely the appearance of concern.

Care carries consequence.

To care is to be affected by another person's wellbeing. It is to notice when something is wrong. It is to give energy that could have been kept for oneself. It is to allow another person's life to matter inside your own.

This is why care cannot be reduced to service.

Service can be automated.

Care must be inhabited.

The future will contain more automated service, more efficient assistance, more responsive systems, and more convenient support.

But human beings will still need care.

Not only the completion of tasks.

The experience of mattering.

A child does not only need food placed before him. He needs the experience of being known.

An aging parent does not only need transportation to an appointment. She needs the dignity of being accompanied.

A friend does not only need an answer. She needs assurance that her pain did not disappear into a system.

A client does not only need a deliverable. He needs confidence that someone understands what is at stake.

Care is the human capacity to recognize the person inside the process.

Without care, systems may become efficient and still feel cruel.

This is why presence matters.

Presence is more than physical proximity.

A distracted person can sit beside someone and still be absent. A present person can make another human being feel less alone in a single moment.

Presence is the gift of undivided attention.

It says, without needing to say it, that the person in front of you is not an interruption.

They are the point.

In an automated world, presence may become more valuable because absence will become easier to disguise.

Messages can be generated. Replies can be scheduled. Images can be posted. Updates can be automated. Relationships can appear maintained from a distance.

Yet the human nervous system often knows the difference.

It knows when attention is real.

It knows when someone is listening.

It knows when another person is emotionally present rather than merely responsive.

This difference will matter.

The future will not suffer from a lack of communication.

It may suffer from a lack of presence.

Forgiveness also remains necessary.

No technology eliminates human rupture.

People will still misunderstand one another. Disappoint one another. Fail one another. Speak carelessly. Withdraw. Act from fear. Choose poorly.

Human beings will still need repair.

Automation can help organize life, but it cannot remove the moral complexity of being human with other humans.

Forgiveness is part of that complexity.

It is not pretending nothing happened. It is not abandoning boundaries. It is not allowing harm to continue.

Forgiveness is the human capacity to make repair possible where repair still belongs.

It allows relationship to survive imperfection.

This is one of the ways human beings differ from systems.

Systems often treat error as failure.

Humans can treat error as a place for repair.

Not always.

Not automatically.

Not without discernment.

But sometimes.

A friendship deepens after repair. A family becomes more honest after rupture. A team becomes stronger after conflict is named and resolved. A person becomes more trustworthy after taking responsibility for harm.

These transformations require forgiveness in some form.

They require people willing to remain engaged with one another after imperfection appears.

A purely optimized world may struggle with this.

Optimization often prefers replacement.

If something fails, replace it.

If someone disappoints, discard them.

If a relationship becomes difficult, exit it.

If a person becomes inconvenient, move on.

Sometimes release is necessary.

But a life built only on replacement becomes shallow.

Nothing matures.

Nothing repairs.

Nothing deepens through difficulty.

Forgiveness protects the possibility of depth.

Loyalty protects it as well.

Loyalty is often misunderstood.

It is not blind allegiance. It is not obedience. It is not remaining inside systems that destroy dignity.

True loyalty is commitment guided by discernment.

It is the decision not to abandon what matters merely because it becomes inconvenient.

Loyalty says that some bonds deserve protection.

Some promises deserve endurance.

Some people deserve patience.

Some responsibilities cannot be automated, delegated, or optimized away.

The future may make loyalty harder because choice will become easier.

There will always be another option. Another platform. Another source. Another person. Another tool. Another community. Another identity. Another life to imagine.

Abundance creates mobility.

Mobility creates freedom.

But too much mobility can weaken commitment.

If every difficulty becomes a reason to leave, nothing has time to become meaningful.

The relational human understands that meaning often requires staying long enough for depth to form.

This does not mean staying everywhere.

It means knowing what deserves loyalty.

A child deserves loyalty.

A worthy friendship deserves loyalty.

A body of work deserves loyalty.

A community may deserve loyalty.

A truth deserves loyalty.

A promise may deserve loyalty.

Human beings become more human through the commitments they choose to keep.

Automation cannot keep a promise for us.

It can remind us. It can assist us. It can reduce the burdens surrounding the promise.

But the promise itself remains human.

The future will therefore still need relational humans.

People who can care when care is inefficient.

People who can remain present when distraction is easier.

People who can forgive where repair is possible.

People who can stay loyal to what deserves endurance.

These capacities may appear soft inside a technological age.

They are not.

They are structural.

Families depend on them.

Communities depend on them.

Companies depend on them.

Schools depend on them.

Civilization depends on them.

A society can become technologically advanced and relationally impoverished. It can automate services while losing care. Increase communication while losing presence. Expand options while losing loyalty. Accelerate everything while losing the patience required for forgiveness.

That is not progress.

That is imbalance.

The Human Advantage Operating System treats relationship as core infrastructure for human life.

Not decoration.

Not sentiment.

Infrastructure.

Because people do not flourish only through efficiency, intelligence, or output.

They flourish through bonds.

They flourish when they are known.

When they are cared for.

When repair is possible.

When loyalty creates enough safety for depth.

The future may automate many forms of assistance. It may generate many forms of interaction. It may simulate many forms of companionship.

But the deepest human need will remain.

To belong to someone.

To matter to someone.

To be met by another human being whose presence is real.

The relational human remains necessary because life is not only a sequence of tasks.

It is a web of attachments.

And no matter how advanced the machines become, human beings will still need other human beings who know how to stay.

Part V

Building the Human Advantage OS

Chapter 23

The Human Advantage Audit

A system cannot be strengthened until it can be seen.

This is true of businesses. It is true of families. It is true of institutions. It is true of a life.

Most people sense when something in their internal operating system is strained, but they do not always know where the strain lives. They feel scattered, reactive, mistrustful, overextended, disconnected, automated, or unreal. They know the world is changing. They know the pace is increasing. They know intelligence is becoming easier to access and labor is becoming easier to automate.

They know the old advantages are weakening.

What they often do not know is which human capacity requires attention first.

That is the purpose of the Human Advantage Audit.

The audit is not a test. It is not a scorecard for proving worth. It is not a productivity exercise disguised as self-knowledge.

It is a way of looking honestly at the human capacities that remain valuable when machines become more capable.

Those capacities are judgment, taste, trust, meaning, relationship, responsibility, attention, and authenticity.

Together they form the Human Advantage Stack.

These capacities are not separate talents operating in isolation. They interact continuously. Judgment depends on attention. Trust depends on responsibility. Taste depends on authenticity. Meaning depends on relationship. Attention protects every other capacity. Authenticity gives the entire system weight.

A person may be strong in one area and underdeveloped in another.

A founder may possess exceptional judgment but weak attention because the system constantly interrupts her. A creator may have strong taste but depleted energy. A professional may be responsible but disconnected from meaning. A leader may be efficient but no longer trusted. A parent may care deeply yet struggle to remain fully present.

The audit helps reveal where the system is strong and where it is leaking.

It begins with judgment.

Judgment asks whether a person can decide what matters when options multiply. This capacity becomes essential in a world of abundant intelligence because artificial intelligence can generate more possibilities than any human can pursue.

A person without judgment becomes overwhelmed by options. They chase every idea, react to every recommendation, and confuse possibility with direction.

A person with judgment understands that not every option deserves a decision.

They ask different questions.

What matters here?

What can be ignored?

What consequence am I not seeing?

What question should I be asking?

Where is the real risk?

What should not be automated?

Judgment is strong when decisions become clearer under pressure.

Judgment is weak when abundance produces paralysis, overreaction, or drift.

The second capacity is taste.

Taste asks whether a person can recognize quality, coherence, and fit when creation becomes easy. Artificial intelligence can generate content, images, ideas, strategies, and variations endlessly. The person without taste mistakes output for significance. They keep more than they should. They publish more than they mean. They pursue ideas simply because they are available.

A person with taste recognizes what deserves to exist.

They ask:

Does this belong?

Does this carry meaning?

Is this coherent with the larger body of work?

Is this merely impressive, or is it alive?

What should be removed?

Taste is strong when selection improves the work.

Taste is weak when more becomes the default answer.

The third capacity is trust.

Trust asks whether others can rely on the person, the work, the judgment, and the pattern over time.

In a world of synthetic abundance, trust becomes a filter. People cannot verify everything. They look for individuals, organizations, and relationships that have earned belief.

Trust is not built through appearance.

It is built through repeated alignment between words and behavior.

A person can evaluate trust by asking:

Do people know what to expect from me?

Do I tell the truth when it is inconvenient?

Do I deliver on what I promise?

Do I acknowledge uncertainty rather than perform certainty?

Do my actions reduce friction for others or increase it?

Trust is strong when reliability compounds.

Trust is weak when people must constantly verify, monitor, interpret, or protect themselves.

The fourth capacity is meaning.

Meaning asks whether a person knows what deserves attention, effort, sacrifice, and loyalty.

Intelligence can answer questions.

Meaning decides which questions matter.

Without meaning, life becomes reactive. A person moves from task to task, goal to goal, trend to trend, tool to tool, without a center.

A person with meaning understands the deeper why beneath the work.

They ask:

What is this for?

What kind of life am I building?

What deserves my energy?

What am I serving?

What would make this success feel empty?

What would make this effort worthwhile?

Meaning is strong when choices align with purpose.

Meaning is weak when achievement increases but coherence declines.

The fifth capacity is relationship.

Relationship asks whether a person is able to belong, care, repair, and remain present with others.

Automation can perform tasks.

It cannot replace belonging.

A person may become highly productive and still feel relationally impoverished. A company may become efficient and still lose trust. A family may be well-managed and still feel emotionally absent.

Relationship requires attention, vulnerability, continuity, and care.

A person can ask:

Who actually receives my presence?

Where have I substituted logistics for connection?

Where is repair needed?

Who depends on me emotionally, not just functionally?

What relationships deserve more time than efficiency allows?

Relationship is strong when belonging deepens.

Relationship is weak when communication exists but connection does not.

The sixth capacity is responsibility.

Responsibility asks whether a person remains accountable for the systems, decisions, tools, and consequences they authorize.

Machines can recommend.

Humans remain accountable.

The automated age will constantly tempt people to hide behind systems. The model suggested it. The algorithm selected it. The process approved it. The tool generated it.

But responsibility cannot be outsourced.

A person can ask:

Can I stand behind this decision?

Who is affected by this outcome?

What am I permitting by default?

Where am I using complexity to avoid accountability?

What needs human review before action is taken?

Responsibility is strong when authority and accountability remain connected.

Responsibility is weak when consequences become diffused across tools, processes, or excuses.

The seventh capacity is attention.

Attention asks whether a person can sustain focus in an environment designed to fragment it.

This may become one of the rarest future skills because intelligence is expanding while human attention remains finite.

Without attention, judgment weakens. Taste weakens. Trust weakens. Meaning weakens. Relationship weakens.

A person can ask:

What receives my best attention?

What fragments me?

Can I remain with an idea long enough to understand it?

Can I listen without preparing my next response?

Do I protect space for depth?

Attention is strong when focus is governed intentionally.

Attention is weak when the mind is constantly captured by whatever arrives next.

The eighth capacity is authenticity.

Authenticity asks whether a person's life, work, relationships, and expression remain rooted in reality.

When synthetic experiences multiply, reality becomes more valuable. A person can increasingly generate appearances without substance. A brand can appear coherent before trust exists. A voice can be imitated before experience is lived. A life can be displayed without being inhabited.

Authenticity requires contact with reality.

A person can ask:

What have I actually lived?

What have I earned through experience?

Where am I performing an image rather than inhabiting a life?

What is real here?

What would remain if the surface disappeared?

Authenticity is strong when outward expression is grounded in lived substance.

Authenticity is weak when appearance grows faster than reality.

Together, these eight capacities create a map.

Not a map of what machines can do.

A map of what humans must strengthen.

The audit works best when answered honestly rather than aspirationally. A person might rate each capacity from one to five. One means the capacity is neglected. Three means it is present but inconsistent. Five means it is strong, visible, and actively practiced.

The number matters less than the pattern.

A low attention score may explain weak judgment. A low meaning score may explain overproduction. A low trust score may explain why opportunities require excessive effort. A low authenticity score may explain why success feels hollow. A low relationship score may explain why life feels efficient but lonely.

The goal is not to perfect every category.

The goal is to identify the next place of human strengthening.

Machines will continue improving. Tools will continue expanding. Automation will continue reaching new domains.

The human response cannot be panic.

It must be development.

The Human Advantage Audit begins that development by asking a simple question:

Where is the human operating system strong, and where does it need repair?

The answer will differ for every person.

But the need to ask will not.

In an age of abundant intelligence and automated labor, the most important upgrade may not be technical.

It may be human.

Human Advantage Audit Wheel

Eight capacities scored around a circle to reveal strengths and gaps.

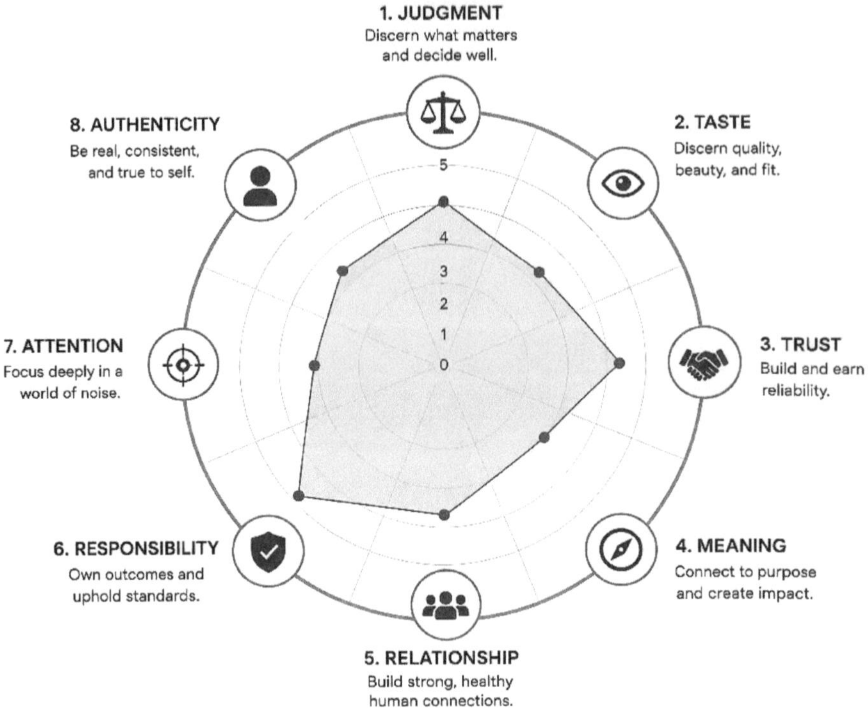

SCORE GUIDE

0 = Very Weak
1 = Weak
2 = Developing
3 = Solid
4 = Strong
5 = Exceptional

HOW TO USE

Score yourself honestly. Focus on your lowest scores to grow with impact.

Chapter 24

Designing a Life Machines Cannot Replace

Most people ask the wrong question.

How do I avoid being replaced?

The question is understandable, especially during periods of rapid technological change. Yet it is also limiting because it assumes the goal is survival. Adapt enough. Learn enough. Automate enough. Stay ahead long enough. Remain useful long enough.

The mindset is defensive.

It frames technology as a threat to escape rather than a reality to understand.

A better question exists.

How do I build a life that becomes more valuable as machines become more capable?

The difference matters.

One approach is built on fear.

The other is built on design.

This chapter is about design.

Not designing a career.

Not designing a business.

Designing a life.

Careers change. Industries change. Technologies change. Tools change. The human operating system remains.

If the previous chapters are correct, the future rewards capacities that machines do not make abundant. Judgment. Taste. Trust. Meaning. Relationship. Responsibility. Attention. Authenticity.

These become the scarce assets.

The challenge is that most people were not taught to build around them.

They were taught to build around tasks.

Acquire a skill.

Perform the task.

Earn the credential.

Get the job.

Advance the role.

Increase the income.

That model worked reasonably well when tasks remained stable.

But the future is not organized around stable tasks.

It is organized around changing tools and enduring human capacities.

This requires a different design philosophy.

Instead of asking, *What should I do?* ask, *Who am I becoming?*

The question shifts from function to formation.

The most durable advantage may not be a specific skill. It may be a specific kind of person. The person who can exercise judgment under uncertainty. The person who can recognize quality. The person who can build trust. The person who can sustain attention. The person who can create meaning. The person who can remain authentic. The person who can care. The person who can assume responsibility.

These capacities travel across industries, technologies, decades, and economic transitions.

A life built around them becomes more resilient.

This is the first principle.

Build around human capacities rather than temporary tasks.

The second principle is more difficult.

Protect what machines cannot give you.

Technology can generate information, but it cannot give you wisdom.
Technology can generate options, but it cannot give you conviction.
Technology can generate entertainment, but it cannot give you belonging.
Technology can generate content, but it cannot give you purpose.

Technology can generate efficiency.

It cannot give you a meaningful life.

Those things must be cultivated intentionally.

This requires a different relationship with time.

Modern systems often treat time as something to maximize. Fill the calendar. Increase utilization. Eliminate waste. Compress delay. Accelerate output.

But a life machines cannot replace contains activities that appear inefficient.

Reading slowly.

Walking.

Thinking.

Observing.

Practicing.

Listening.

Gardening.

Creating.

Resting.

Having long conversations.

Watching children grow.

Building friendships.

Sitting beside people during difficult seasons.

These activities often generate little measurable output.

Yet they create enormous human value.

A person who removes them entirely may become productive while becoming impoverished.

The third principle is to build depth instead of volume.

Artificial intelligence will produce volume. Automation will produce volume. Algorithms will produce volume. Scale will become increasingly common.

Depth remains scarce.

A deep friendship.

A deep body of work.

A deep understanding.

A deep commitment.

A deep expertise.

A deep relationship with place.

A deep relationship with purpose.

Depth takes time.

It cannot be rushed. It cannot be mass-produced. It cannot be copied instantly.

The future may reward depth precisely because everything else accelerates.

This applies to work as much as life.

A person may publish thousands of pieces of content. The deeper question is whether any of them matter. A person may generate endless ideas. The deeper question is whether any deserve pursuit. A person may accumulate followers. The deeper question is whether trust exists.

Depth creates durability.

Volume creates visibility.

The future will contain plenty of visibility.

Depth may become the greater advantage.

The fourth principle is to build a life that remains meaningful without constant optimization.

This may be one of the hardest lessons of the automated age.

Optimization is addictive. There is always another improvement, another efficiency, another productivity system, another automation, another tool, another metric, another shortcut.

The process never ends.

At some point, optimization must serve life rather than become life.

A person should be able to answer a simple question:

If every productivity tool disappeared tomorrow, what would still matter?

The answer reveals the architecture of the life.

The child.

The spouse.

The friend.

The book.

The garden.

The work.

The community.

The conversation.

The craft.

The purpose.

These are often the things that survive technological change.

Not because they are old.

Because they are human.

The fifth principle is to remain connected to reality.

The future will contain increasingly convincing simulations. Synthetic experiences. Synthetic content. Synthetic relationships. Synthetic identities. Synthetic environments.

Many will be useful.

Some will be extraordinary.

But a life machines cannot replace remains anchored in reality.

Real people.

Real conversations.

Real places.

Real commitments.

Real experiences.

Real consequences.

Reality creates perspective. Reality creates humility. Reality creates wisdom. Reality creates meaning.

A person who loses contact with reality eventually loses contact with themselves.

The sixth principle is to become irreplaceable to specific people rather than impressive to everyone.

Many people spend enormous energy trying to become broadly valuable. More visible. More recognized. More admired. More influential.

Yet some of the most meaningful forms of value are highly specific.

A child does not need the world's best parent.

The child needs their parent.

A friend does not need universal expertise.

The friend needs someone who cares.

A community does not need endless content.

It needs trustworthy people.

The future may reward people who understand this distinction.

Not because influence is unimportant.

Because relationship is deeper.

Machines may become impressive.

Humans remain meaningful.

The seventh principle is to design for stewardship rather than extraction.

Many systems extract. They extract attention, energy, time, emotion, labor, and resources.

A life machines cannot replace is built differently.

It asks:

What am I protecting?

What am I cultivating?

What am I building that outlasts the moment?

What am I responsible for?

What am I passing forward?

Stewardship creates continuity. It connects present choices to future consequences. It transforms life from consumption into contribution.

This is one reason purpose becomes increasingly important as automation expands.

The machine can optimize.

The human must decide what is worth optimizing.

The machine can produce.

The human must decide what deserves production.

The machine can accelerate.

The human must decide where to go.

This leads to the final principle.

Build around what remains scarce.

Not what remains fashionable.

Not what remains trendy.

Not what remains profitable for the moment.

What remains scarce.

Trust remains scarce.

Attention remains scarce.

Authenticity remains scarce.

Meaning remains scarce.

Care remains scarce.

Wisdom remains scarce.

Judgment remains scarce.

The future will contain more intelligence than any previous generation imagined.

But intelligence alone does not create a life.

A life is created through choices, commitments, relationships, experiences, responsibilities, stories, values, and presence.

These things remain profoundly human.

This is why the goal is not to compete with machines.

It is to become more fully human than the systems around you require.

A machine may write.

You can still originate.

A machine may recommend.

You can still judge.

A machine may generate content.

You can still create meaning.

A machine may simulate connection.

You can still belong.

A machine may optimize the process.

You can still decide the purpose.

The future will reward many forms of intelligence.

But the deepest forms of human value may emerge from a simpler question.

What kind of life would still matter even if every machine became more capable tomorrow?

The answer to that question is the beginning of a life machines cannot replace.

Chapter 25

Designing a Business Machines Cannot Replace

For generations, businesses competed through execution.

The company that could manufacture faster often won. The company that could process more transactions often won. The company that could answer more calls, ship more products, complete more work, and scale more efficiently often gained the advantage.

Execution mattered because execution was scarce.

Many businesses were designed around this reality. They hired people to perform tasks, built processes around those tasks, measured output from those tasks, and created value through increasingly efficient execution. The organizations that reduced friction, standardized operations, and improved throughput often outperformed competitors.

Artificial intelligence and automation are changing that equation.

Execution is becoming increasingly abundant. Not perfect. Not universal. But increasingly available. Machines can generate reports, review documents, analyze information, draft content, answer questions, route requests, and organize workflows. Robotics is beginning to extend that same capability into the physical world.

As execution becomes easier to obtain, companies face a different challenge.

What remains valuable when everyone can execute?

This may be the central question of the post-automation business.

The answer is not technology alone. Technology rarely creates durable advantage because technology eventually spreads. What one company adopts today, another company often adopts tomorrow. The tool that feels unique eventually becomes expected.

As execution becomes commoditized, long-term advantage moves elsewhere.

It moves toward what remains scarce.

Trust.

Judgment.

Taste.

Relationship.

Meaning.

Reputation.

Community.

Identity.

Purpose.

The future business may begin to look less like a machine and more like a human operating system.

At first, this sounds strange. Businesses have traditionally been taught to become more systematic, more scalable, and less dependent on individual people. Those lessons remain valuable. But they are incomplete.

A company may automate nearly every process and still become replaceable.

Because process alone does not create attachment.

People rarely remain loyal to a process. They remain loyal to trust. To experience. To relationships. To meaning. To identity. To outcomes they cannot easily find elsewhere.

This is why designing a business machines cannot replace begins with a different question.

Not:

What can be automated?

But:

What should remain human?

The distinction matters.

Many organizations approach technology by asking how much labor can be removed. A stronger approach asks where human judgment creates disproportionate value. Where does trust matter? Where does interpretation matter? Where does relationship matter? Where does care matter? Where does accountability matter?

These are the areas where human value compounds rather than erodes.

Consider a professional services firm. Artificial intelligence may review documents, summarize contracts, draft communications, analyze risk, and generate recommendations. All of these capabilities create efficiency.

Yet clients rarely hire trusted advisors solely because they can produce documents.

They hire them because they trust judgment.

The document becomes easier to generate.

The judgment becomes more valuable.

The same pattern appears in healthcare. Technology can improve diagnostics, monitor conditions, identify patterns, and analyze data. Yet patients often remain loyal to physicians who communicate clearly, exercise judgment, demonstrate care, and create trust.

The future doctor may use extraordinary tools.

The relationship remains human.

The same pattern appears in education. Artificial intelligence can explain concepts, personalize learning, generate lessons, and answer questions. Yet students still benefit from teachers who recognize potential, build confidence, provide encouragement, and understand context.

The future classroom may become more technological.

The development of human beings remains human.

This creates a useful design principle:

Automate the process.

Protect the relationship.

Automate the routine.

Protect the judgment.

Automate the workflow.

Protect the trust.

Automate the repetition.

Protect the meaning.

Many organizations will get this backward.

They will automate what differentiates them and preserve what does not. They will reduce human interaction because it appears expensive, only to discover that human interaction was the source of loyalty. They will optimize customer service until service becomes sterile. They will automate communication until communication loses trust.

Eventually they will pursue efficiency until they become indistinguishable from competitors using the same tools.

This is one of the hidden risks of automation.

When every company has access to similar technology, differentiation becomes harder. Technology levels the playing field.

Human qualities create separation.

This is where culture becomes important.

Culture is often treated as a secondary consideration—something soft, internal, or difficult to measure. In reality, culture may become one of the strongest competitive advantages of the automated age.

Culture influences decisions, standards, behavior, trust, accountability, innovation, and adaptability. It determines how people behave when no procedure exists.

Machines handle predictable situations.

Culture governs unpredictable ones.

And the future will continue producing unpredictability.

A company with strong culture adapts.

A company without culture depends entirely on process.

Process works until conditions change.

Culture helps people navigate the change.

The same principle applies to reputation.

In a world of synthetic content and abundant production, reputation becomes increasingly valuable. A trusted company gains leverage not because it has better marketing, but because people believe it will behave consistently.

Trust reduces friction.

Clients move faster.

Partners cooperate more easily.

Employees commit more deeply.

Opportunities arrive more readily.

The company becomes easier to work with because uncertainty declines.

This creates a compounding effect.

The future business therefore accumulates trust as deliberately as it accumulates capital.

Both are assets.

One appears on the balance sheet.

The other often does not.

Yet both shape outcomes.

Another important principle involves community.

Many traditional businesses focused primarily on transactions. A product exchanged for money. A service exchanged for money. A task exchanged for money.

Transactions remain important.

But communities create resilience.

A transaction solves a momentary problem.

A community creates ongoing connection.

Communities form around shared values, shared experiences, shared interests, and shared identity. Technology can support community, but it struggles to create genuine community on its own.

This is why some businesses become movements while others remain commodities.

The movement creates belonging.

Belonging is difficult to replace.

The commodity competes primarily on price.

Price is easy to replace.

The future business therefore benefits from understanding the difference.

The goal is not merely delivering value.

The goal is creating a relationship with the people receiving value.

The strongest businesses of the future may not simply sell products.

They may curate worlds.

Worlds of trust.

Worlds of meaning.

Worlds of identity.

Worlds of participation.

People do not only buy what a business produces.

They buy what a business represents.

This is where purpose becomes important.

Purpose is frequently misunderstood as branding, mission statements, or marketing language. In reality, purpose answers a practical question:

Why does this organization deserve to exist?

The question becomes more important as production becomes easier.

Why does this product exist?

Why does this service exist?

Why does this company exist?

What human need does it serve?

What value does it protect?

What future does it create?

Companies that answer these questions clearly often become more resilient because they are built around meaning rather than mechanics.

Mechanics change.

Meaning endures.

This does not eliminate the importance of operational excellence. Execution still matters. Technology still matters. Automation still matters.

But they become supporting systems rather than the entire strategy.

The company of the future may resemble a layered structure. Automation handles routine execution. Artificial intelligence supports analysis. Systems manage workflow. Humans provide judgment. Culture provides alignment. Trust provides leverage. Relationships provide resilience. Purpose provides direction.

Together these elements create a business that is difficult to replace.

Not because it resists technology.

Because it uses technology to strengthen what remains uniquely human.

The goal is not to build a company that competes with machines.

Machines will continue improving.

The goal is to build a company that becomes more valuable because machines improve.

A company where automation removes friction.

A company where technology amplifies judgment.

A company where efficiency creates space for deeper human value.

A company where trust grows faster than output.

A company where relationships become assets.

A company where meaning survives scale.

A company where people remain central even when systems become powerful.

This is the operating system of the post-automation business.

Not a company designed around tasks.

A company designed around human advantage.

Because execution is becoming abundant.

The businesses that endure will be the ones that understand what remains scarce—and build around it.

Chapter 26

Designing Education for the Human Advantage

For more than a century, education has largely been preparing children for a world that no longer exists.

This is not a criticism of teachers. Nor is it a criticism of schools. It is the predictable consequence of rapid change. Educational systems are designed to create stability. Technological revolutions create instability.

The two often move at different speeds.

For generations, education focused on preparing students for a world where information was scarce, expertise was concentrated, and work was organized around predictable tasks. Students learned facts, memorized procedures, followed instructions, completed assignments, demonstrated competency, and accumulated credentials.

The model made sense.

Employers needed workers who could reliably execute tasks. Institutions needed people who could operate inside established systems. Knowledge itself carried value because access to knowledge was limited. The student who knew more possessed an advantage.

The world rewarded information.

The world now rewards something different.

Information is increasingly available. Intelligence is increasingly available. Answers are increasingly available. Artificial intelligence can explain concepts, summarize books, generate lesson plans, solve equations, write essays, answer questions, translate languages, and create educational content in seconds.

This changes the educational landscape.

Not because learning becomes unnecessary.

Because learning changes.

The question is no longer, *How do we help children access information?*

The question becomes, *How do we help children develop capacities that remain valuable when information is abundant?*

This may be one of the most important educational questions of the century.

The answer begins with a difficult realization.

Many of the skills most heavily rewarded by traditional education are becoming easier to automate. Memorization, retrieval, routine analysis, standardized production, and procedural execution still matter, but they are no longer enough.

The future belongs to children who can do more than produce answers.

The future belongs to children who can ask better questions.

This requires discernment.

Discernment is the ability to determine what matters, what deserves attention, what deserves skepticism, what deserves pursuit, and what deserves rejection. Artificial intelligence can provide answers. A child must still decide whether the answer is meaningful, correct, complete, ethical, useful, and aligned with reality.

This capacity cannot be downloaded.

It must be developed.

The future student therefore requires more than knowledge.

The future student requires judgment.

A child growing up today will likely encounter more information before adulthood than previous generations encountered in entire lifetimes. Without discernment, abundance becomes confusion. With discernment, abundance becomes opportunity.

This is why education must increasingly teach students how to think rather than simply what to think.

The distinction matters.

A person who knows many facts may still become overwhelmed. A person who understands how to evaluate information can continue learning indefinitely.

The second capacity is creativity.

For years, creativity was often treated as secondary. Important, perhaps, but less practical. The serious work was mathematics, science, writing, testing, and measurement. Creativity occupied the margins.

The future reverses part of this equation.

Not because mathematics or science become less important.

Because creativity becomes more important.

When machines can generate information, the ability to originate becomes valuable. Children need opportunities to imagine, experiment, build, question, design, create, fail, revise, and explore.

Creativity is not an extracurricular activity.

It is a survival skill.

The future will continuously present novel situations. New technologies. New ethical dilemmas. New social systems. New economic realities. Old answers will not always fit.

Creative people generate new ones.

This requires educational environments where curiosity is protected rather than suppressed, where questions are welcomed rather than merely answered, where experimentation matters as much as correctness, and where students learn that creating something original often begins with uncertainty.

The third capacity is embodied experience.

This may become increasingly important as life becomes increasingly digital.

A child cannot develop entirely through screens.

Screens can teach. They can explain. They can connect. They can support. But they cannot replace lived experience.

A child needs physical reality. Nature. Movement. Exploration. Conversation. Observation. Responsibility.

The physical world teaches lessons differently than information does.

A child learns patience by growing a garden. Responsibility by caring for an animal. Courage by trying something difficult. Resilience by failing and recovering. Wonder by exploring something real.

These experiences shape judgment because they involve consequence.

Reality provides feedback.

Not theoretical feedback.

Actual feedback.

The future may contain extraordinary simulations.

Children will still need reality.

The same principle applies to relationships.

Human beings develop socially through interaction, conflict, friendship, cooperation, misunderstanding, repair, negotiation, and belonging. These experiences cannot be fully replaced by information.

They must be lived.

Education therefore cannot become purely informational.

It must remain developmental.

The goal is not merely transferring knowledge.

The goal is helping children become capable human beings.

The fourth capacity is courage.

This may be the most overlooked educational objective.

Knowledge without courage accomplishes little. A person may know what is right and still fail to act. A person may understand the opportunity and still remain afraid. A person may see the truth and still remain silent.

The future requires courage because uncertainty is increasing.

There is no stable roadmap for the coming decades. Careers will change. Industries will change. Technologies will change. Social systems will change.

Children will need the confidence to enter situations where the answers do not already exist.

This confidence does not emerge from certainty.

It emerges from experience.

A child develops courage by doing difficult things. Trying. Failing. Recovering. Trying again.

The process matters.

Educational systems often focus heavily on correctness.

Yet courage develops through imperfection. Through risk. Through uncertainty. Through effort.

A child who never struggles often becomes fragile.

A child who learns to struggle productively becomes resilient.

Resilience may become one of the most valuable future assets.

This leads to a larger realization.

The purpose of education is changing.

For generations, education prepared children for jobs.

The future still requires employment.

But education can no longer stop there.

The future requires people capable of governing intelligence rather than merely consuming it. People capable of exercising judgment, creating, building relationships, discerning truth, sustaining attention, carrying responsibility, and living meaningful lives.

These capacities extend beyond employment. They shape citizenship, leadership, parenting, friendship, community, and human flourishing.

The future educational question therefore becomes surprisingly simple.

Not, *What information should children learn?*

But, *What kind of people should children become?*

The answer matters because technology increasingly changes what humans can do.

Education increasingly determines who humans become.

The Human Advantage Operating System suggests that schools, parents, communities, and educational systems should orient around the capacities machines cannot easily provide: judgment, creativity, embodied experience, courage, attention, meaning, relationship, responsibility, and authenticity.

These capacities may become more valuable with each technological advance.

Not because technology fails.

Because technology succeeds.

The more intelligence becomes available, the more important discernment becomes. The more automation expands, the more important purpose becomes. The more simulation improves, the more important reality becomes. The more answers become available, the more important wisdom becomes.

This is why the future of education is not primarily about competing with machines.

It is about helping children become more fully human.

The goal is not to raise children who can outperform artificial intelligence at its strengths.

The goal is to raise children who develop strengths artificial intelligence does not make abundant.

Children who can see clearly.

Think deeply.

Create courageously.

Act responsibly.

Love loyally.

Pay attention.

Tell the truth.

And build meaningful lives in a world overflowing with capability.

That may be the most important educational objective of all.

Chapter 27

The Future Belongs to the More Human

For a long time, the future was imagined as a race.

Humans against machines. Workers against automation. Artists against algorithms. Teachers against software. Drivers against autonomous vehicles. Writers against artificial intelligence.

The image was simple.

The machine advances.

The human tries to stay ahead.

Learn faster.

Adapt faster.

Produce faster.

Automate faster.

Upgrade faster.

Do not fall behind.

Do not become obsolete.

Do not become replaceable.

The metaphor is powerful because fear likes speed. When people are afraid, they want movement. They want a strategy. They want protection. They want to know what to learn, what to buy, what to automate, what to avoid, and where to run next.

But the metaphor is wrong.

The goal is not to outrun machines.

Machines will continue getting faster. They will calculate faster, generate faster, translate faster, sort faster, summarize faster, move faster, respond faster, and produce faster.

A human being designed around speed alone will eventually lose to systems built for acceleration.

The human advantage cannot be found in becoming a weaker version of the machine.

It must be found somewhere else.

The goal is not to outrun machines.

The goal is to stop abandoning what machines cannot become.

That is the central idea of this book.

The future does not belong to the most automated human.

It belongs to the more human one.

The person who is more discerning, more responsible, more trustworthy, more creative, more attentive, more relational, more authentic, and more grounded in meaning.

This is not sentimental.

It is strategic.

Every technological era shifts value toward what remains scarce. If intelligence becomes abundant, wisdom becomes valuable. If content becomes abundant, taste becomes valuable. If synthetic experience becomes abundant, reality becomes valuable. If automation becomes abundant, responsibility becomes valuable. If connection becomes constant, presence becomes valuable.

The future belongs to those who understand this new scarcity map.

Not those who panic at every new capability.

Not those who deny the capability exists.

Not those who worship the machine.

Not those who reject it.

But those who recognize that technology changes the location of human value.

This recognition creates a different posture.

Less fear.

More discernment.

Less chasing.

More building.

Less imitation.

More origination.

Less exhaustion.

More stewardship.

A person does not become more valuable by automating every corner of life. A business does not become more valuable by removing every human layer. A school does not become more valuable by turning learning into machine-assisted compliance. A society does not become more valuable by treating human beings as inefficient obstacles.

The question is not how much can be automated.

The question is what automation is for.

If automation returns time, what receives the time?

If intelligence returns answers, what questions deserve attention?

If machines remove labor, what human capacities are strengthened?

If systems produce more, what meaning governs the production?

The future will test whether human beings know how to answer these questions.

This test will appear quietly.

Not only in laboratories, boardrooms, or policy debates.

It will appear in ordinary life.

A parent choosing whether to hand a child another screen or sit beside the child in conversation. A founder choosing whether to automate customer contact or deepen trust. A teacher choosing whether to outsource thinking or cultivate discernment. A creator choosing whether to generate endlessly or say something true.

A professional choosing whether to hide behind a recommendation or take responsibility.

A person choosing whether to fill every recovered minute or reclaim life.

These choices may seem small.

They are not.

They are the operating system.

The future is not built only through major inventions.

It is built through repeated decisions about what is allowed to matter.

The danger of advanced technology is not simply that machines become powerful.

The danger is that humans become passive.

Passive consumers.

Passive workers.

Passive learners.

Passive citizens.

Passive creators.

Passive recipients of recommendations, predictions, prompts, nudges, and generated realities.

Passivity is one of the great risks of abundance.

When everything is provided, the human capacity to choose can weaken. When everything is answered, the human capacity to question can weaken. When everything is optimized, the human capacity to endure meaningful difficulty can weaken. When everything is generated, the human capacity to originate can weaken.

This is why becoming more human requires active practice.

Judgment must be practiced.

Taste must be practiced.

Trust must be earned.

Meaning must be chosen.

Relationship must be tended.

Responsibility must be carried.

Attention must be protected.

Authenticity must be lived.

None of these capacities emerge automatically.

They develop through use.

A person becomes discerning by making choices under uncertainty. A person develops taste by paying attention to quality. A person becomes trustworthy by acting consistently over time. A person finds meaning through commitment. A person deepens relationships through presence and repair. A person becomes responsible by owning consequences.

A person strengthens attention by resisting fragmentation.

A person becomes authentic by remaining in contact with reality.

These are human forms of training.

They are not obsolete.

They may become essential.

The future belongs to the more human because machines cannot relieve human beings of the responsibility to become.

A machine can assist with learning.

It cannot become wise for a child.

A machine can support a business.

It cannot become trustworthy for a founder.

A machine can generate a song.

It cannot live the longing that makes music necessary.

A machine can simulate conversation.

It cannot offer the mortal presence of another person who has chosen to stay.

A machine can recommend action.

It cannot carry human accountability.

The boundary is not always clean.

Technology will continue entering domains once believed exclusively human. It will surprise us. It will unsettle us. It will perform tasks we assumed would remain protected. It will imitate behaviors we associate with intelligence, creativity, care, and companionship.

That is why the Human Advantage Operating System cannot rest on denial.

It must rest on depth.

The more machines can imitate the surface, the more humans must cultivate the source.

The source of judgment.

The source of taste.

The source of trust.

The source of meaning.

The source of relationship.

The source of responsibility.

The source of attention.

The source of authenticity.

The surface can be copied more easily than the source can be lived.

This is why a real life matters.

Experience matters.

Memory matters.

Embodiment matters.

Place matters.

Mortality matters.

Care matters.

Commitment matters.

A human being is not merely a processor of information.

A human being is a life unfolding through time.

That life carries particularity.

A specific childhood.

Specific relationships.

Specific losses.

Specific loyalties.

Specific risks.

Specific recoveries.

Specific loves.

Specific questions.

Specific ways of seeing.

From this particularity comes human originality.

Not originality as novelty.

Originality as lived perspective.

The future may contain infinite generated perspectives.

But the human perspective rooted in actual life remains different.

It has weight.

It has consequence.

It has memory.

It has skin in the world.

That difference matters.

It may matter even more as synthetic reality expands.

The future therefore asks for a particular kind of courage.

Not the courage to fight machines.

The courage to remain human in systems that reward machine-like behavior.

The courage to be slow where slowness creates meaning.

The courage to be present where distraction is easier.

The courage to take responsibility where blame can be diffused.

The courage to build trust where performance would be faster.

The courage to create from truth rather than merely generate from pattern.

The courage to protect attention when every system wants to capture it.

This courage will not always be applauded.

Sometimes it will look inefficient.

Sometimes it will look old-fashioned.

Sometimes it will look unscalable.

Sometimes it will look like saying no to what everyone else is adopting.

But human advantage does not mean using every tool everywhere.

It means knowing what belongs.

Technology should serve the human operating system.

Not replace it.

That is the final principle.

Use intelligence.

Do not surrender judgment.

Use automation.

Do not surrender responsibility.

Use generation.

Do not surrender taste.

Use connection.

Do not surrender relationship.

Use efficiency.

Do not surrender meaning.

Use simulation.

Do not surrender reality.

Use the future.

Do not abandon the human.

The coming age will likely be extraordinary. It will contain capabilities that previous generations could not imagine. It will solve problems, create new ones, expand access, disrupt professions, rearrange power, and force societies to reconsider what work, education, creativity, and life are for.

No one can fully predict it.

But one thing is already visible.

The human future will not be secured by becoming less human.

It will be secured by becoming more deliberately human.

More conscious of what we are giving away.

More protective of what cannot be manufactured.

More careful with energy, attention, trust, and meaning.

More willing to ask whether progress is serving life.

The future belongs to the more human.

Not because machines will remain weak.

Because machines will become strong.

And the stronger they become, the more clearly we will see what they still cannot become.

That is where the human advantage remains.

Not in speed.

Not in endless output.

Not in imitation.

But in the capacities that make life worth living and systems worth building.

The goal is not to outrun machines.

The goal is to stop abandoning what machines cannot become.

That is the Human Advantage Operating System.

And it begins again every time a human being chooses to remain fully present, fully responsible, and fully alive.

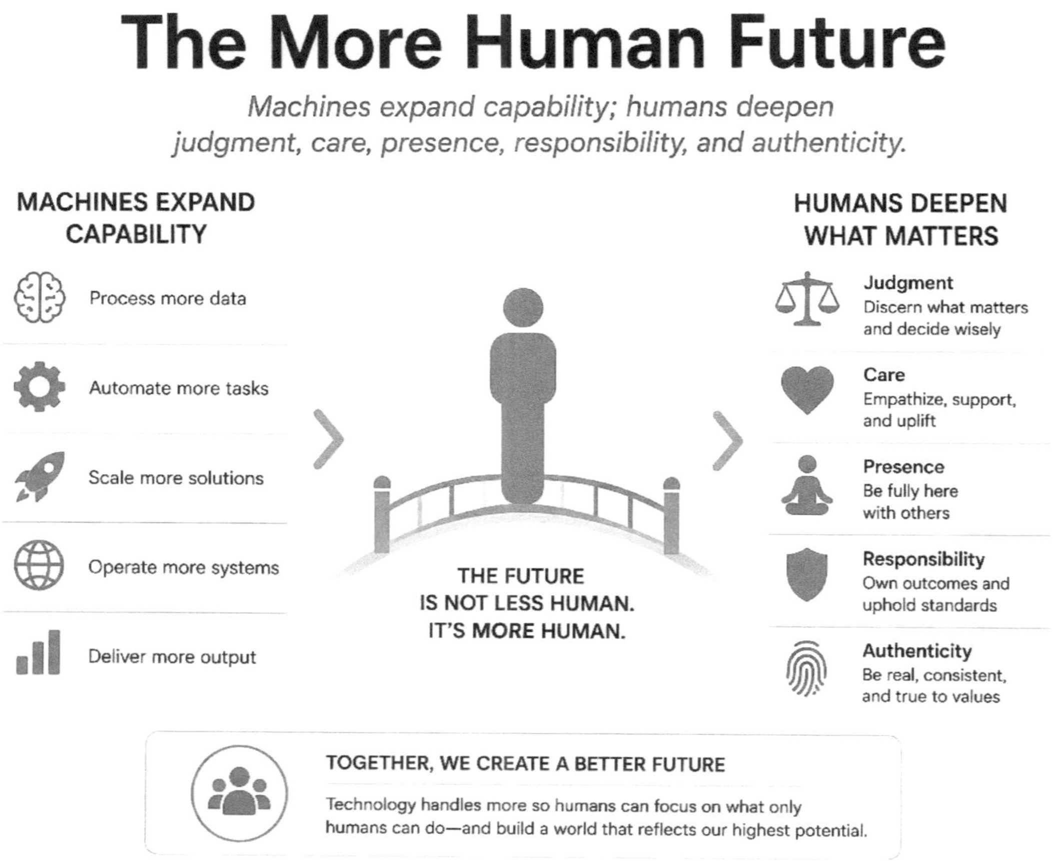
The More Human Future
Machines expand capability; humans deepen
judgment, care, presence, responsibility, and authenticity.
MACHINES EXPAND CAPABILITY
Process more data
Automate more tasks
Scale more solutions
Operate more systems
Deliver more output
THE FUTURE
IS NOT LESS HUMAN.
IT'S MORE HUMAN.
HUMANS DEEPEN WHAT MATTERS
Judgment
Discern what matters and decide wisely
Care
Empathize, support, and uplift
Presence
Be fully here with others
Responsibility
Own outcomes and uphold standards
Authenticity
Be real, consistent, and true to values
TOGETHER, WE CREATE A BETTER FUTURE
Technology handles more so humans can focus on what only humans can do—and build a world that reflects our highest potential.

Conclusion

The Human Advantage

The future will not ask less of human beings.

It will ask something different.

For a long time, people assumed that progress would gradually make human effort less necessary. Machines would take over labor. Computers would take over calculation. Software would take over coordination. Artificial intelligence would take over analysis. Robotics would take over physical work.

In one sense, that story is true.

Many forms of work will become easier. Many tasks will become faster. Many processes will become automated. Many answers will become instantly available.

But ease is not the same as wisdom.

Speed is not the same as direction.

Output is not the same as meaning.

Automation does not eliminate the human question.

It reveals it.

What are we for?

That question sits beneath every technological revolution. It becomes louder each time machines become more capable. As intelligence becomes abundant, human beings must learn to govern intelligence. As labor becomes automated, human beings must decide what recovered life is for. As creation becomes infinite, human beings must recognize what deserves to exist. As synthetic experience expands, human beings must remain connected to reality.

The human advantage is not found in pretending machines are weak.

It is found in understanding where machines are strong and refusing to surrender what strength alone cannot produce.

This book has argued that certain capacities become more valuable precisely because technology improves.

Judgment.

Taste.

Trust.

Meaning.

Relationship.

Responsibility.

Attention.

Authenticity.

These are not sentimental leftovers from an older world.

They are the operating capacities of a human future.

The more powerful the tools become, the more important these capacities become. The future will reward those who can decide what matters when options multiply. Those who can recognize quality when creation becomes easy. Those who can earn trust when appearances can be generated. Those who can create meaning when answers are everywhere.

It will reward those who can sustain relationships when interaction becomes automated, accept responsibility when recommendations come from systems, protect attention when everything competes for it, and remain authentic when simulation becomes convincing.

This is the human advantage.

Not superiority over machines.

Not denial of machines.

Not fear of machines.

A different form of strength.

The strength to remain human in an age that will constantly tempt people to become more mechanical.

The strength to pause when speed is praised. To choose when options multiply. To care when efficiency would be easier. To tell the truth when performance would be smoother. To create from lived experience rather than endless generation.

To take responsibility when blame could be diffused.

To stay present when distraction is everywhere.

To build lives, businesses, schools, families, and communities around what remains real.

The future belongs to the more human.

Not because technology will stop.

Because technology will not stop.

And the more it advances, the more clearly we will see the difference between capability and wisdom.

Machines may become extraordinary. They may help cure disease, accelerate discovery, reduce drudgery, expand education, support creativity, and open possibilities previous generations could not imagine.

They may also create confusion, dependence, imitation, overload, distrust, and systems that move faster than human judgment can comfortably govern.

The difference will not be determined by the machines alone.

It will be determined by the human operating systems surrounding them.

By parents deciding how children will learn.

By founders deciding what businesses should become.

By creators deciding what deserves expression.

By teachers deciding what deserves attention.

By leaders deciding what should remain human.

By citizens deciding what kind of future they are willing to build.

The future is ultimately shaped by people who use tools without being used by them.

People who understand that the point of progress is not endless acceleration.

The point is a life more fully inhabited.

A business more deeply trusted.

An education more alive.

A community more connected.

A future more worthy of the humans who must live inside it.

This book began with a question:

What remains valuable when intelligence becomes abundant and labor becomes automated?

The answer is not a list of jobs.

It is not a prediction.

It is not a defensive strategy.

The answer is a set of human capacities that become more important as technology improves.

The answer is an invitation.

An invitation to strengthen judgment when answers become cheap.

To cultivate taste when creation becomes infinite.

To earn trust when appearances become easy.

To choose meaning when options multiply.

To protect relationships when interaction becomes automated.

To carry responsibility when systems become powerful.

To reclaim attention when distraction becomes profitable.

To remain authentic when simulation becomes convincing.

The future will not be secured by becoming less human.

It will be built by those willing to become more human on purpose.

That is the Human Advantage Operating System.

And it begins wherever a person decides that the parts of life machines cannot become are the parts most worth protecting.

Because the deepest human advantage was never speed.

It was never efficiency.

It was never output alone.

The deepest human advantage is the ability to choose what matters, commit to it, care for it, and build a life around it.

Everything else in this book flows from that truth.

And every generation must decide whether it will be forgotten or lived.

Author Reflection

Where Do We All Fit In?

I did not come to this book as a technologist trying to predict the future.

I came to it as a mother, a founder, a writer, a lawyer, a creator, and a person who has lived long enough to recognize when the ground beneath a system begins to move.

Throughout my life, I have spent time inside many different systems. Legal systems. Business systems. Family systems. Educational systems. Creative systems. Financial systems. Social systems.

I have watched how people behave inside those systems. I have watched how systems protect people and how they fail them. I have watched how they create clarity and how they create confusion. I have watched people become trapped inside processes that were originally designed to serve them.

That is why I tend to think in operating systems.

Not because life is mechanical.

Because life is patterned.

And once a pattern becomes visible, it can be understood.

This book began with a question that felt larger than artificial intelligence alone:

What remains valuable when intelligence becomes abundant and labor becomes automated?

But underneath that question was another one.

A more personal one.

Where do we all fit?

Where does the parent fit?

Where does the child fit?

Where does the founder fit?

Where does the artist fit?

Where does the teacher fit?

Where does the aging person fit?

Where does the worker fit?

Where does the person who still wants a meaningful life fit?

The answer is not simple.

But it is not hopeless.

We fit where human life still carries weight.

We fit where judgment is required. We fit where trust must be earned. We fit where someone must care. We fit where meaning must be chosen. We fit where responsibility cannot be outsourced. We fit where attention must be protected. We fit where reality still matters.

We fit wherever presence changes the outcome.

This realization feels deeply personal to me.

As a mother, I do not want my children prepared only to compete with machines. I want them prepared to become fully human in a world where machines will be increasingly capable.

I want them to know how to think.

How to notice.

How to ask better questions.

How to create.

How to recover.

How to tell the truth.

How to love people.

How to use tools without becoming tools.

Those capacities feel more important to me than any specific technical skill because the tools will continue changing. The human qualities underneath them must endure.

As a founder, I do not want to build a business that survives only because it executes tasks faster than someone else.

I want to build something trusted.

Something clear.

Something useful.

Something governed by judgment rather than efficiency alone.

Technology can improve execution. It can accelerate workflows. It can remove friction. But trust, responsibility, and meaning remain human work.

Those are the parts I want to strengthen.

As a writer, I do not want to produce words simply because words can now be produced quickly.

I want to originate.

I want to notice what is happening beneath the surface and give language to it. I want to explore questions that matter before they become obvious. I want to create books that help people see something they had not seen before.

For me, writing has never been about volume.

It has always been about recognition.

As a person, I do not want the future to make me less human.

I want it to return me to what matters.

That may be the quiet hope hidden inside this entire book.

The future does not have to erase us.

But it will force us to become more conscious about what we refuse to surrender.

We cannot surrender judgment to convenience.

We cannot surrender taste to volume.

We cannot surrender trust to appearances.

We cannot surrender meaning to optimization.

We cannot surrender relationship to interaction.

We cannot surrender responsibility to systems.

We cannot surrender attention to noise.

We cannot surrender authenticity to simulation.

These are not abstract ideas.

They are daily decisions.

This is where we all fit.

Not outside the future.

Inside it.

But not as passive recipients.

As stewards.

As parents.

As creators.

As founders.

As teachers.

As friends.

As citizens.

As people who understand that technology should serve life rather than replace the reasons life matters.

The Human Advantage Operating System is not a rejection of machines.

It is a reminder of the human.

Because the more powerful machines become, the more important it becomes to remember what they are for.

They are not here to tell us what matters.

That remains our work.

They are not here to choose our values.

That remains our work.

They are not here to decide what deserves loyalty, sacrifice, attention, or love.

That remains our work.

Perhaps that is where we fit best.

Not in a race against machines.

Not in a competition to become faster, more efficient, or more optimized than the systems we create.

But in the responsibility to become the kind of humans who can wisely live with them.

The future will almost certainly contain extraordinary intelligence, extraordinary automation, and extraordinary capability.

What I hope it also contains is extraordinary humanity.

People willing to think deeply.

People willing to care.

People willing to tell the truth.

People willing to create.

People willing to remain present.

People willing to take responsibility.

People willing to protect what cannot be manufactured.

Because in the end, the question is not whether machines become more capable.

The question is whether human beings remember what they are capable of becoming.

That is the question I continue asking.

And perhaps that is where we all fit.

In the ongoing work of becoming more fully human while the future unfolds around us.

A machine can generate output.

A human leaves a mark.

APPENDICES

Appendix A

The Human Advantage Stack

THE EIGHT CAPACITIES THAT BECOME MORE VALUABLE AS INTELLIGENCE BECOMES ABUNDANT

Every era rewards a different set of capabilities.

Agricultural societies rewarded physical labor and land stewardship. Industrial societies rewarded production and efficiency. Information societies rewarded knowledge and expertise.

The emerging age of artificial intelligence and automation rewards something different.

As intelligence becomes increasingly accessible and execution becomes increasingly automated, value migrates toward capacities that remain scarce. The Human Advantage Stack identifies eight such capacities.

These are not temporary trends.

They are enduring human advantages.

Machines may support them. Machines may amplify them. Machines may interact with them. But machines do not make them abundant.

For this reason, they become increasingly valuable.

LEVEL 1: JUDGMENT
THE ABILITY TO DETERMINE WHAT MATTERS

Artificial intelligence can generate options.

Judgment determines which option deserves action.

Artificial intelligence can analyze information. Judgment determines which information is significant. Artificial intelligence can recommend. Judgment decides.

Judgment functions as the governance layer above intelligence. Without judgment, abundant intelligence becomes abundant confusion.

A person with strong judgment understands that not every possibility deserves pursuit. They know how to identify hidden risks, competing priorities, unintended consequences, and the questions that matter more than the answers.

Judgment becomes increasingly valuable because abundance creates complexity. The more options available, the more important it becomes to choose wisely.

Key questions include:

- What matters most?
- What can be ignored?
- What risk is hidden?
- What consequence is being overlooked?
- What deserves attention now?

The human advantage is the ability to choose wisely when possibilities multiply.

Level 2: Taste
The Ability to Recognize Quality

When creation becomes abundant, selection becomes valuable.

Taste is the ability to recognize quality, coherence, fit, beauty, significance, and alignment. It is not merely preference. Taste is discernment applied to creation.

A person with taste recognizes what belongs. A person without taste accumulates options.

This distinction becomes increasingly important in a world where machines can generate content endlessly. When everyone can create, the ability to recognize what deserves existence becomes more valuable than the ability to produce more of it.

Taste helps people identify what strengthens the whole and what weakens it. It helps distinguish what is meaningful from what is merely impressive.

Key questions include:

- Does this belong?
- Does this strengthen the whole?
- Is this meaningful or merely impressive?
- What should be removed?

The human advantage is the ability to recognize what deserves to exist.

Level 3: Trust
The Ability to Earn Belief

In a world overflowing with information, trust becomes a filtering system.

People cannot verify everything. They increasingly rely on trusted individuals, organizations, creators, and communities to help them navigate complexity.

Trust is earned through consistency, integrity, transparency, and reliability over time. It reduces friction, accelerates cooperation, and compounds through repeated alignment between words and actions.

Trust is not built through appearance.

It is built through behavior.

As synthetic content expands and information becomes harder to evaluate, trusted sources become increasingly valuable.

Key questions include:

- Can people rely on me?
- Do my actions align with my words?
- Do I tell the truth when it is inconvenient?
- Am I consistent over time?

The human advantage is the ability to become a trusted source in a world of uncertainty.

LEVEL 4: MEANING
THE ABILITY TO DETERMINE WHAT IS WORTH PURSUING

Intelligence answers questions.

Meaning determines which questions matter.

Meaning organizes attention, effort, sacrifice, loyalty, and purpose. It provides direction to capability and significance to achievement.

Without meaning, achievement becomes empty.

Without meaning, efficiency loses direction.

Meaning functions as the compass of the human operating system. It answers the deeper question beneath success, productivity, and ambition.

Key questions include:

- What is this for?
- What deserves my life?
- What matters most?
- What kind of person am I becoming?

The human advantage is the ability to orient a life toward significance.

Level 5: Relationship
The Ability to Belong

Automation can perform tasks.

It cannot replace belonging.

Human beings remain relational creatures. We require connection, care, loyalty, repair, forgiveness, and presence. Relationships create resilience, identity, meaning, and support.

A person may become highly productive and still feel disconnected. A company may become efficient and still lose trust. A family may be well-managed and still feel emotionally absent.

Relationship reminds us that human beings do not flourish alone.

Key questions include:

- Who receives my presence?
- Who depends on me?
- Where is repair needed?
- What relationships deserve deeper investment?

The human advantage is the ability to create genuine human connection.

Level 6: Responsibility
The Ability to Own Consequences

Machines can recommend.

Humans remain accountable.

Responsibility connects authority to consequence. It requires ownership rather than avoidance and integrity rather than excuse.

As automation expands, people will increasingly be tempted to hide behind systems, algorithms, and recommendations.

Responsibility resists that temptation.

It asks who remains accountable when a decision affects real people.

Key questions include:

- Can I stand behind this decision?
- Who is affected?
- What consequence am I accepting?
- What responsibility belongs to me?

The human advantage is the willingness to own outcomes rather than hide behind systems.

Level 7: Attention
The Ability to Focus

Information is abundant.

Attention remains scarce.

Attention determines what enters the human operating system. It protects every other capacity in the stack. Without attention, judgment weakens. Meaning weakens. Relationships weaken. Trust weakens.

The ability to sustain focus may become one of the rarest future skills because intelligence is expanding while human attention remains finite.

Attention is not merely awareness.

It is intentional focus.

Key questions include:

- What receives my best attention?
- What fragments my focus?
- What deserves deeper concentration?
- What should be excluded?

The human advantage is the ability to remain present in a distracted world.

LEVEL 8: AUTHENTICITY
THE ABILITY TO REMAIN CONNECTED TO REALITY

As synthetic experiences expand, reality becomes increasingly valuable.

Authenticity emerges from lived experience. Real relationships. Real commitments. Real places. Real memories. Real consequences.

Authenticity is not performance.

It is alignment between life and expression.

As appearances become easier to manufacture, authenticity becomes more valuable because it remains rooted in reality rather than simulation.

Key questions include:

- What is real here?
- What have I actually lived?
- Where am I performing instead of participating?
- What remains if appearances disappear?

The human advantage is the ability to remain grounded in reality when simulation becomes abundant.

How the Stack Works

The Human Advantage Stack is not a checklist.

It is an operating system.

Each capacity strengthens the others. Judgment improves taste. Taste improves trust. Trust strengthens relationships. Relationships deepen meaning. Meaning guides responsibility. Responsibility sharpens attention. Attention protects authenticity. Authenticity reinforces judgment.

The system becomes self-reinforcing.

Likewise, weakness in one area often affects the others. Fragmented attention weakens judgment. Weak judgment weakens responsibility. Weak responsibility weakens trust. Weak trust weakens relationships. Weak relationships weaken meaning.

The stack functions as a living system rather than a collection of isolated traits.

The Central Principle

The future will contain extraordinary intelligence.

The future will contain extraordinary automation.

The future will contain extraordinary efficiency.

The Human Advantage Stack identifies what remains scarce.

Not because machines are incapable.

Because machines do not make these qualities abundant.

The goal is therefore not to compete with machines at their strengths.

The goal is to strengthen the capacities that become more valuable as machines improve.

Judgment.

Taste.

Trust.

Meaning.

Relationship.

Responsibility.

Attention.

Authenticity.

Together they form the Human Advantage Operating System.

Together they answer the central question of this book:

What remains valuable when intelligence becomes abundant and labor becomes automated?

The answer is not less human.

The answer is more human.

Appendix B

The New Scarcity Map

HOW VALUE MIGRATES IN THE AGE OF ARTIFICIAL INTELLIGENCE

Every economic era is governed by scarcity.

People often assume value comes from usefulness alone. Usefulness certainly matters, but usefulness by itself does not create value. Air is useful. Sunlight is useful. Water is useful. Yet value emerges when something useful is also scarce.

Scarcity shapes markets. It shapes institutions, education, status, incentives, and behavior. In many ways, the history of civilization can be understood as a series of changing scarcity maps.

At different points in history, different resources became difficult to obtain. Land was scarce. Labor was scarce. Capital was scarce. Energy was scarce. Information was scarce. Expertise was scarce. Access was scarce. Distribution was scarce.

Each era rewarded the people who possessed what others could not easily obtain.

As technology evolves, scarcity moves.

And when scarcity moves, value moves with it.

This appendix provides a map of that movement. It is not a prediction of every future outcome. It is a framework for understanding where human value is likely to migrate as intelligence becomes abundant and labor becomes increasingly automated.

The Old Scarcity Map

For much of the industrial and information ages, value accumulated around a familiar set of scarcities.

Information was one of them.

People with access to information possessed leverage. Libraries, universities, experts, consultants, researchers, and institutions all existed partly because knowledge was difficult to obtain. Information itself created advantage.

Analysis was another scarcity.

Having information was not enough. Someone had to interpret it, compare it, organize it, synthesize it, and transform it into decisions. Analysis required expertise and time. Because it was difficult, it carried value.

Production was scarce as well.

Books required publishers. Products required factories. Music required studios. Media required distribution networks. Creation itself was constrained by physical and economic limitations.

Labor also remained scarce.

Most systems depended directly on human effort. Tasks required people. Processes required people. Movement required people. Labor created value because labor was necessary.

Access mattered because gatekeepers controlled opportunity. Editors, managers, universities, networks, and institutions determined who entered important systems.

Distribution mattered because even valuable ideas required a way to reach other people. The ability to distribute often mattered as much as the ability to create.

This was the old scarcity map.

Much of modern society was built around it.

The Transition

Artificial intelligence and automation do not eliminate value.

They relocate value.

This is the critical insight.

Technology makes some things abundant. Abundance reduces scarcity. Reduced scarcity shifts value elsewhere.

The future therefore requires a new map.

The question is no longer simply what machines can do.

The question is what becomes scarce when machines can do more.

From Information to Judgment

Information becomes abundant.

Judgment becomes scarce.

For centuries, possessing information created advantage. Today information is increasingly available to everyone. Answers can be retrieved instantly. Explanations can be generated on demand.

The challenge is no longer access.

The challenge is deciding what matters.

Judgment determines what deserves attention, what deserves action, what can be ignored, and what consequence is being overlooked.

When everyone has access to answers, judgment becomes valuable.

FROM ANALYSIS TO DISCERNMENT

Artificial intelligence can analyze.

Humans must discern.

Analysis processes information. Discernment governs information.

Discernment asks different questions. Which conclusions matter? Which assumptions are flawed? Which risks deserve attention? Which recommendations belong inside reality?

As analysis becomes easier, discernment becomes more valuable.

FROM PRODUCTION TO TASTE

Creation becomes abundant.

Taste becomes scarce.

Artificial intelligence can generate content, products, strategies, images, music, and designs at extraordinary scale. The challenge shifts from creation to selection.

Taste is the ability to recognize quality, coherence, significance, beauty, and fit. It determines what belongs and what should be removed.

When everything can be generated, selection becomes power.

FROM LABOR TO RESPONSIBILITY

Automation performs tasks.

Humans remain accountable.

As systems become more capable, responsibility becomes increasingly valuable because someone must still answer for consequences.

Who approved this?

Who owns the risk?

Who carries the outcome?

Who remains accountable when something goes wrong?

Capability expands.

Responsibility becomes scarce.

From Access to Trust

Access becomes easier.

Trust becomes harder.

People can reach almost anyone. Information can travel almost anywhere. Communication barriers continue shrinking.

The question is no longer whether someone can be reached.

The question is whether someone deserves belief.

Trust becomes a filtering system. It becomes infrastructure. It becomes leverage.

From Distribution to Reputation

Anyone can publish.

Anyone can create.

Anyone can distribute.

The scarcity shifts toward credibility.

Reputation becomes increasingly valuable because it reduces uncertainty. In a world overflowing with content, trusted sources become anchors.

People no longer ask only, *Can I find information?*

They increasingly ask, *Who stands behind it?*

From Content to Meaning

Content becomes infinite.

Meaning remains scarce.

The future rewards people who can answer questions that content alone cannot answer.

Why does this matter?

What is this for?

What deserves commitment?

Content informs.

Meaning directs.

As production expands, significance becomes more valuable.

From Connection to Relationship

Technology creates connection.

Humans still require belonging.

The scarcity shifts from communication to relationship.

The question is no longer whether people can interact. The question is whether people can care, trust, remain present, repair, and belong.

Connection becomes abundant.

Relationship remains scarce.

From Stimulation to Attention

Content expands.

Attention remains finite.

The future rewards people who can sustain focus in a world designed to fragment it. Information is abundant. Stimulation is abundant. Distraction is abundant.

Attention remains scarce.

The ability to govern attention becomes increasingly valuable.

From Simulation to Authenticity

Synthetic experiences multiply.

Reality becomes valuable.

As simulations become more convincing, authenticity becomes more important. People increasingly seek what is real, lived, embodied, and earned.

Authenticity connects human beings to experience, presence, truth, and reality.

When simulation expands, authenticity gains value.

The Human Advantage Zone

Notice something important about the new scarcity map.

Nearly every item is deeply human.

Judgment.

Discernment.

Taste.

Trust.

Meaning.

Relationship.

Responsibility.

Attention.

Authenticity.

These are not primarily technological capabilities.

They are human capacities.

Technology may support them. Technology may amplify them. Technology may strengthen them.

But technology does not make them abundant.

This is why the future belongs to people who strengthen these capacities.

The advantage shifts.

Not toward competing with machines.

Toward becoming more capable of what machines do not make common.

The Strategic Implication

Most people respond to technological change by asking:

What skill should I learn?

The better question is:

Which human capacity is becoming more valuable?

Skills change.

Tools change.

Industries change.

Scarcity reveals where value is moving.

The New Scarcity Map suggests that long-term advantage comes from developing stronger judgment, deeper discernment, better taste, greater trustworthiness, stronger relationships, clearer meaning, deeper responsibility, more disciplined attention, and greater authenticity.

These capacities travel across industries, technologies, economic transitions, and generations.

They remain valuable because they remain scarce.

THE FINAL OBSERVATION

Every technological revolution changes what humans do.

The artificial intelligence revolution changes what humans should become.

The old scarcity map rewarded information.

The new scarcity map rewards wisdom.

The old scarcity map rewarded production.

The new scarcity map rewards significance.

The old scarcity map rewarded access.

The new scarcity map rewards trust.

The old scarcity map rewarded output.

The new scarcity map rewards meaning.

This is the central pattern of the Human Advantage Operating System.

Technology does not eliminate value.

Technology relocates value.

The future belongs to those who understand where it moved.

Appendix C

Human Advantage Audit

Human Advantage Audit Wheel

Eight capacities scored around a circle to reveal strengths and gaps.

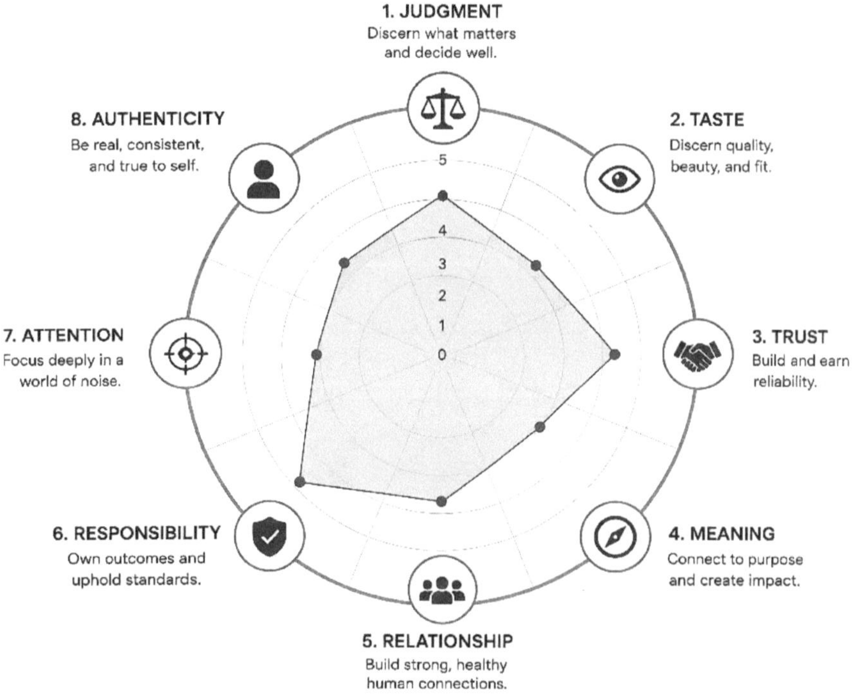

SCORE GUIDE

0 = Very Weak
1 = Weak
2 = Developing
3 = Solid
4 = Strong
5 = Exceptional

HOW TO USE

Score yourself honestly. Focus on your lowest scores to grow with impact.

Assessing the Human Operating System

Most people periodically assess their finances.

They review accounts.

Assets.

Liabilities.

Income.

Debt.

Risk.

The process provides visibility.

Without visibility, improvement becomes difficult.

The same principle applies to the human operating system.

Artificial intelligence and automation are changing what creates value. Many traditional advantages are becoming easier to access. Information is abundant. Analysis is abundant. Content is abundant. Execution is increasingly abundant.

The question is no longer simply:

What can I do?

The question becomes:

What human capacities am I developing?

This audit provides a framework for assessing the eight capacities that form the Human Advantage Stack:

1. Judgment

2. Taste

3. Trust

4. Meaning

5. Relationship
6. Responsibility
7. Attention
8. Authenticity

The purpose is not perfection.

The purpose is awareness.

The audit identifies strengths, weaknesses, opportunities for development, and areas where the human operating system may require strengthening.

Scoring Instructions

For each statement, assign a score:

1 = Rarely True

2 = Occasionally True

3 = Sometimes True

4 = Usually True

5 = Consistently True

Answer honestly.

Not aspirationally.

The goal is clarity.

---

1. Judgment Audit
The Ability to Determine What Matters

___ I can identify the most important issue in a complex situation.

___ I avoid becoming overwhelmed when presented with many options.

___ I recognize second-order consequences before making decisions.

___ I know when additional information will not improve a decision.

___ I distinguish urgency from importance.

___ I can identify what should be ignored.

___ I make decisions without requiring complete certainty.

___ I recognize patterns that others miss.

___ I know when technology should assist and when human judgment should intervene.

___ I remain calm when ambiguity increases.

Judgment Score: ______ / 50

---

2. Taste Audit
The Ability to Recognize Quality

___ I can distinguish quality from popularity.

___ I recognize when something belongs and when it does not.

___ I remove unnecessary complexity rather than adding more.

___ I recognize coherence across ideas, projects, and decisions.

___ I value significance over novelty.

___ I notice details that affect quality.

___ I know when enough is enough.

___ I can identify meaningful opportunities from distracting opportunities.

___ I regularly refine rather than simply accumulate.

___ I trust my informed perception even before consensus forms.

Taste Score: ______ / 50

3. Trust Audit
The Ability to Earn Belief

___ People know what to expect from me.

___ I do what I say I will do.

___ I tell the truth even when it creates discomfort.

___ I acknowledge uncertainty when I do not know something.

___ My behavior remains consistent under pressure.

___ Others rely on me during difficult situations.

___ I address mistakes directly rather than hiding them.

___ My actions align with my stated values.

___ I build long-term credibility rather than short-term appearances.

___ People trust me with responsibility.

Trust Score: ______ / 50

---

4. Meaning Audit
The Ability to Know What Matters

___ I understand what I am building my life around.

___ My daily actions connect to larger purposes.

___ I know what deserves my time and energy.

___ I can explain why my work matters.

___ I understand the difference between achievement and meaning.

___ I make decisions according to values rather than trends.

___ I know what I am unwilling to sacrifice.

___ I regularly reflect on purpose.

___ My goals align with my deeper priorities.

___ I understand what a successful life means to me.

Meaning Score: ______ / 50

---

5. Relationship Audit
The Ability to Belong and Connect

___ I make time for important relationships.

___ People feel heard when speaking with me.

___ I am emotionally present during conversations.

___ I address conflict rather than avoiding it indefinitely.

___ I practice forgiveness where appropriate.

___ I invest in long-term relationships.

___ I maintain meaningful friendships.

___ People know they matter to me.

___ I create space for belonging.

___ I prioritize people over convenience when appropriate.

Relationship Score: ______ / 50

---

6. Responsibility Audit
The Ability to Own Consequences

___ I accept responsibility for my decisions.

___ I do not hide behind systems, tools, or circumstances.

___ I consider how my actions affect others.

___ I acknowledge mistakes quickly.

___ I address problems directly.

___ I understand the consequences of my choices.

___ I can stand behind difficult decisions.

___ I avoid shifting blame unnecessarily.

___ I follow through on commitments.

___ I act consistently with my obligations.

Responsibility Score: ______ / 50

---

7. Attention Audit
The Ability to Sustain Focus

___ I can focus deeply without constant interruption.

___ I protect time for meaningful work.

___ I avoid unnecessary distractions.

___ I can remain present during conversations.

___ I finish important tasks before switching attention.

___ I deliberately choose what receives my attention.

___ I consume information intentionally.

___ I spend time reflecting rather than constantly reacting.

___ I maintain focus despite digital distractions.

___ I regularly create space for deep thinking.

Attention Score: ______ / 50

---

8. Authenticity Audit
The Ability to Remain Connected to Reality

___ My public image reflects my actual life.

___ I do not perform an identity I do not inhabit.

___ My values appear consistently across different settings.

___ I maintain contact with real experiences rather than only representations.

___ I spend meaningful time in physical reality.

___ I know the difference between appearance and substance.

___ I am honest with myself about strengths and weaknesses.

___ I prioritize lived experience over image management.

___ I seek truth even when it challenges my preferences.

___ My life feels aligned rather than fragmented.

Authenticity Score: ______ / 50

Human Advantage Summary

Capacity	Score
Judgment	______
Taste	______
Trust	______
Meaning	______
Relationship	______
Responsibility	______

Capacity	Score
Attention	______
Authenticity	______

Total Score: ______ / 400

Interpreting Results

320–400

Highly Developed Human Advantage System

You possess strong foundations across most dimensions. Continue strengthening weaker areas while protecting existing strengths.

240–319

Capable but Uneven

Several capacities are strong, but weaknesses in one area may limit the whole system. Focus on the lowest scoring category first.

160–239

Significant Development Opportunity

The human operating system requires deliberate strengthening. Identify one or two capacities to improve over the next year.

Below 160

Reactive Rather Than Designed

The system may be driven more by circumstance than intentional development. Begin by focusing on attention, meaning, and responsibility.

The Growth Principle

The purpose of this audit is not comparison.

The purpose is direction.

Machines will continue improving.

Tools will continue expanding.

Automation will continue accelerating.

The most important question is not whether technology is advancing.

It is whether the human operating system is advancing as well.

The future rewards people who strengthen the capacities technology does not make abundant.

Judgment.

Taste.

Trust.

Meaning.

Relationship.

Responsibility.

Attention.

Authenticity.

These capacities form the Human Advantage Stack.

They are not fixed traits.

They are skills.

They can be practiced.

Developed.

Strengthened.

Refined.

The audit simply reveals where the work begins.

And where the next version of the human advantage can be built.

Appendix D

Work That Remains Valuable

A Practical Guide to Human Work After Automation

Not all work disappears when machines become more capable.

Some work changes form. Some work becomes less scarce. Some work becomes more valuable.

The mistake is assuming that work can be divided into only two categories: work machines can do and work machines cannot do.

Reality is more nuanced.

A better distinction is this: work that is primarily execution becomes easier to automate, while work that requires judgment, meaning, trust, care, responsibility, and lived context remains valuable.

This appendix identifies the kinds of work most likely to retain and even increase their value in an age of abundant intelligence and automated labor.

Work That Requires Judgment

Judgment remains valuable because information alone does not make decisions.

Machines can analyze information, compare alternatives, identify patterns, and generate recommendations. Human beings must still decide what matters.

This category includes strategic decision-making, risk assessment, legal interpretation, medical judgment, ethical review, crisis response, leadership decisions, complex prioritization, exception handling, and ambiguous problem-solving.

The common feature is context.

The answer depends on timing, consequence, tradeoffs, uncertainty, and responsibility.

Machines may assist.

Humans remain accountable for the decision.

Work That Requires Trust

Trust becomes more valuable as synthetic content expands.

People increasingly need reliable humans, brands, institutions, and advisors to help them navigate complexity. Information may be abundant, but confidence remains scarce.

This category includes trusted advisory services, leadership, coaching, counseling, teaching, client relationship management, sensitive negotiations, healthcare relationships, financial guidance, legal guidance, and community leadership.

People do not only need answers.

They need confidence in the person helping them interpret the answers.

That confidence is built through trust.

Work That Requires Taste

When creation becomes abundant, recognition becomes scarce.

Taste is the ability to determine what deserves to exist, what deserves attention, and what deserves investment.

This category includes editing, curation, creative direction, brand direction, product selection, design judgment, publishing decisions, artistic direction, experience design, and portfolio development.

Machines can generate options endlessly.

Humans still recognize quality, coherence, fit, and significance.

The more options that exist, the more valuable taste becomes.

Work That Requires Meaning

Meaning determines why something matters.

Machines can produce content.

Humans create significance.

This category includes writing rooted in lived experience, spiritual and philosophical guidance, purpose-driven leadership, narrative development, legacy work, educational design, cultural interpretation, mission-building, values-based strategy, and community storytelling.

People seek more than information.

They seek orientation.

Meaning helps answer the question beneath the question.

Why does this matter?

Work That Requires Relationship

Automation can perform tasks.

It cannot replace belonging.

Human beings still need care, presence, loyalty, repair, and connection. These needs remain regardless of technological progress.

This category includes parenting, teaching, caregiving, therapy, mentorship, friendship-based community building, team leadership, conflict repair, long-term client service, and relational sales.

The value comes not from task completion.

The value comes from human connection.

Work That Requires Responsibility

Machines can recommend.

Humans remain accountable.

As systems become more capable, responsibility becomes increasingly valuable because someone must still own consequences.

This category includes executive leadership, governance, compliance oversight, legal sign-off, safety review, medical responsibility, fiduciary responsibility, AI governance, quality assurance, and institutional accountability.

Technology can assist.

Responsibility cannot be outsourced.

Someone must still stand behind the decision.

Work That Requires Attention

Sustained attention becomes increasingly rare in a distracted world.

This category includes deep research, long-form writing, complex analysis, serious study, investigation, skilled craft, meditation and reflection practices, deep listening, strategic planning, and creative development.

Depth requires focus.

Machines can assist with information.

They cannot focus a human life on what matters.

As information expands, the ability to sustain attention becomes increasingly valuable.

Work That Requires Authenticity

As synthetic experiences multiply, reality becomes more valuable.

This category includes live performance, physical books, handmade craft, local experiences, personal storytelling, original art, real-world teaching, nature-based work, place-based community building, and human-centered hospitality.

People continue seeking experiences grounded in reality.

Not because technology fails.

Because authenticity remains scarce.

Work That Requires Embodiment

Human beings live in bodies.

Not everything can be reduced to information.

This category includes physical therapy, massage and bodywork, sports training, dance, music performance, skilled trades, personal care, culinary arts, gardening, and hands-on teaching.

Embodied work carries presence, timing, sensation, and lived experience.

The physical world still matters.

That reality does not disappear because intelligence becomes digital.

WORK THAT REQUIRES MORAL COURAGE

Some work remains valuable because it requires a person to stand for something.

This category includes whistleblowing, ethical leadership, advocacy, truth-telling, protecting vulnerable people, resisting harmful systems, making unpopular but necessary decisions, defending standards, saying no, and taking responsibility under pressure.

Machines do not carry conscience.

They can support analysis.

They cannot replace moral courage.

WORK THAT REQUIRES CONTEXT

Many decisions cannot be separated from the surrounding human system.

This category includes organizational diagnosis, family guidance, cultural interpretation, community planning, conflict resolution, negotiation, change management, institutional memory, customer insight, and human systems design.

Context is often invisible to outsiders and difficult to reduce to data.

It lives inside relationships, history, incentives, and human experience.

Understanding context remains valuable because context shapes meaning.

WORK THAT REQUIRES REPAIR

The future will not eliminate conflict, error, loss, misunderstanding, or rupture.

Repair remains human work.

This category includes mediation, conflict resolution, therapy, leadership repair, customer recovery, family repair, institutional reform, community healing, reputation repair, and trust rebuilding.

Human systems break in human ways.

Repair requires truth, responsibility, patience, humility, and care.

Those capacities remain difficult to automate.

The Work That Weakens

Some forms of work become more vulnerable as automation expands.

Work becomes easier to automate when it is primarily repetitive, procedural, rule-based, easily standardized, low-context, low-trust, low-judgment, high-volume, pattern-driven, and detached from human consequence.

This does not mean such work disappears entirely.

It means the scarcity declines.

And when scarcity declines, value often migrates elsewhere.

The Work That Strengthens

Work becomes more valuable when it requires judgment, trust, taste, meaning, relationship, responsibility, attention, authenticity, embodiment, courage, context, and repair.

These are the areas where human advantage remains strongest.

They are valuable not because machines are weak.

They are valuable because machines do not make these qualities abundant.

Practical Reflection

To evaluate your own work, consider the following questions:

What parts of my work are primarily execution?

What parts require judgment?

What parts require trust?

What parts require taste?

What parts require human relationship?

What parts require responsibility?

What parts require lived experience?

What parts would lose value if automated too aggressively?

What human capacity should I strengthen next?

What should I stop identifying with because it is only a task?

The answers often reveal where value is migrating inside your own career.

Final Principle

The future does not eliminate work.

It reveals which work was merely task performance and which work carried human value.

Work remains valuable when it requires a human operating system.

Judgment.

Taste.

Trust.

Meaning.

Relationship.

Responsibility.

Attention.

Authenticity.

The strongest future work will not be the work machines cannot touch at all.

It will be the work humans know how to govern, deepen, humanize, and make meaningful.

That is the work that remains valuable.

Appendix E

Questions for Parents, Founders, Creators, and Leaders

A Reflection Guide for the Human Advantage Age

The future will not be built solely by technology.

It will be built by the choices people make while using technology.

Every era creates new tools. Every era creates new possibilities. Every era creates new forms of power. Yet the deeper question is always the same: How will human beings respond?

The Human Advantage Operating System is ultimately not a book about machines.

It is a book about people.

How they think.

How they choose.

How they lead.

How they raise children.

How they build companies.

How they create.

How they remain human inside systems becoming increasingly powerful.

The questions in this appendix are designed for reflection rather than quick answers. Some may take minutes to answer. Some may take years.

Their purpose is not certainty.

Their purpose is clarity.

Questions for Parents

Children are growing up in a world where information, intelligence, and automation will be increasingly available. The challenge is no longer preparing them to access answers.

The challenge is preparing them to govern answers wisely.

Parents therefore face a different responsibility than previous generations. It is not enough to help children succeed academically. They must also help them develop the human capacities that remain valuable when knowledge becomes abundant.

Consider the following questions:

What human capacities am I helping my child develop?

Can my child distinguish truth from persuasion?

Does my child know how to think, or only how to answer?

Does my child know how to tolerate uncertainty?

Is curiosity being protected or replaced by convenience?

Does my child spend meaningful time in physical reality?

Is my child developing courage through experience?

Does my child know how to form relationships?

Does my child understand responsibility?

What qualities do I hope remain when technology changes?

A useful reflection is this:

If artificial intelligence could answer every academic question my child ever asked, what would still need to be taught?

QUESTIONS FOR FOUNDERS

Automation changes execution.

It does not eliminate leadership.

In many cases, it increases the importance of leadership because technology amplifies whatever direction already exists.

Founders increasingly face decisions about what should be automated, what should remain human, and what truly differentiates their organizations.

Consider:

What part of my business is truly differentiated?

What should be automated?

What should remain human?

Does my company compete on execution or trust?

Does my organization create meaning or merely activity?

What would clients miss if we disappeared tomorrow?

Where does judgment create the greatest value?

What part of the company depends entirely on me?

What part should never depend entirely on me?

Am I building a business or merely operating one?

A useful reflection is this:

If every competitor gained access to the same technology tomorrow, what would still make my company valuable?

Questions for Creators

Creation is becoming easier.

Originality is becoming more valuable.

The future creator is not simply a producer. The future creator is an originator.

The challenge is no longer generating content.

The challenge is generating perspective.

Consider:

What do I notice that others overlook?

What questions continue following me?

What perspective comes from my lived experience?

Am I creating from truth or chasing attention?

Does my work reveal something meaningful?

What would I create if algorithms disappeared?

What deserves expression?

What deserves silence?

What deserves years rather than weeks?

What am I becoming through my creative practice?

A useful reflection is this:

What part of my work could only exist because I lived the life I have lived?

Questions for Leaders

Leadership increasingly involves governing systems rather than merely managing tasks.

Technology expands capability.

Leadership governs consequence.

As tools become more powerful, leaders must think more carefully about what is being optimized and what is being sacrificed.

Consider:

What is being optimized?

What is being sacrificed?

What standards am I protecting?

What trust am I responsible for maintaining?

Where am I relying on systems instead of judgment?

What decisions require human oversight?

What risks are invisible to the dashboard?

What would people say my leadership truly values?

What culture am I creating?

What future am I encouraging others to build?

A useful reflection is this:

If efficiency improved dramatically tomorrow, what would I do with the energy that was returned?

Questions About Judgment

Judgment determines what matters when possibilities multiply.

As information becomes abundant, the ability to determine significance becomes increasingly valuable.

Ask yourself:

What matters most right now?

What am I paying attention to that does not deserve attention?

What am I ignoring that deserves attention?

What difficult truth am I avoiding?

What assumption needs examination?

What consequence am I underestimating?

What should not be automated?

Questions About Taste

Taste determines what deserves existence.

It governs selection in a world overflowing with options.

Ask yourself:

What belongs?

What should be removed?

What creates clarity?

What creates noise?

What deserves existence?

What am I keeping because it is available rather than valuable?

What would improve if it became simpler?

Questions About Trust

Trust is built through repeated alignment between words and behavior.

In a world where appearances become easier to manufacture, trust becomes increasingly important.

Ask yourself:

Who trusts me?

Why?

What behavior created that trust?

Where have I weakened trust?

What promise remains unfinished?

What relationship requires consistency rather than explanation?

What reputation am I building through repeated actions?

Questions About Meaning

Meaning determines what deserves attention, effort, sacrifice, and loyalty.

It provides direction to capability.

Ask yourself:

What is this for?

What deserves my life?

What would make success feel empty?

What am I unwilling to sacrifice?

What future am I building?

What do I hope remains after I am gone?

What truly matters?

Questions About Relationship

Human beings flourish through belonging.

Relationships require attention, care, repair, and presence.

Ask yourself:

Who needs more of my presence?

What relationship deserves repair?

What conversation am I postponing?

Where am I substituting efficiency for connection?

Who knows they matter to me?

Who have I neglected?

What would deeper belonging require?

Questions About Responsibility

Responsibility connects authority to consequence.

It asks whether we remain accountable for the systems and decisions we influence.

Ask yourself:

What belongs to me?

What am I avoiding?

What consequence am I creating?

Where am I hiding behind systems?

What decision requires ownership?

What commitment needs follow-through?

Can I stand behind my choices?

Questions About Attention

Attention determines what enters the human operating system.

Without attention, every other capacity weakens.

Ask yourself:

What receives my best attention?

What captures attention without earning it?

What deserves deeper focus?

What fragments my thinking?

What would improve if I protected my attention more carefully?

Where am I reacting instead of choosing?

What deserves sustained focus?

Questions About Authenticity

Authenticity asks whether life remains connected to reality.

It is the alignment between lived experience and outward expression.

Ask yourself:

What is real here?

What am I performing?

What am I hiding?

What have I actually experienced?

Where does my life align with my values?

Where does it not?

What would remain if appearances disappeared?

The Final Questions

The Human Advantage Operating System ultimately leads to a small number of enduring questions.

Questions that technology cannot answer for us.

Questions every generation must answer again.

Who am I becoming?

What deserves my attention?

What deserves my loyalty?

What deserves my life?

What should remain human?

What should be protected?

What should be built?

What should be passed forward?

These questions matter because the future is not only a technological challenge.

It is a human one.

Artificial intelligence will continue advancing. Automation will continue expanding. Capabilities will continue increasing.

The central question remains unchanged:

Will human beings become more capable without becoming less human?

The answer will not come from machines.

It will come from the choices people make.

Parents.

Founders.

Creators.

Leaders.

Citizens.

Families.

Communities.

The future belongs to those who strengthen the capacities technology does not make abundant.

Judgment.

Taste.

Trust.

Meaning.

Relationship.

Responsibility.

Attention.

Authenticity.

The questions in this appendix are not intended to produce perfect answers.

They are intended to keep those capacities alive.

Because the quality of the future may depend less on what machines become and more on what humans choose to remain.

About the Author

Lisa J. Scott writes at the intersection of systems, meaning, and human transformation. Her work explores how people, families, businesses, institutions, and societies operate beneath the surface—and what becomes possible when those operating systems are made visible.

She is the founder of in2edge, a woman-owned legal services and business operations company supporting complex contract, transition, and carve-out work for private equity, corporate, and legal teams. Her professional background includes law, business, language, global experience, and systems design.

As an author, Lisa creates books that combine clarity, structure, reflection, and insight. Her Operating System series examines the hidden architecture of modern life, including creativity, aging, freedom, business, conflict, leadership, judgment, and human advantage. Across her work, she asks a consistent question: What is really happening underneath the visible event?

Lisa is also a mother, creator, and founder of Lisa J. Scott Publishing. Her books reflect a belief that the future will not be shaped only by technology, power, or efficiency, but by the human capacities that remain scarce: judgment, taste, trust, meaning, relationship, responsibility, attention, and authenticity.

She lives in Texas, where she continues to build a body of work around books as worlds, systems as language, and the human advantage as something worth protecting.

Books by Lisa J. Scott

A living body of work

The Core Trilogy

The Woman from Frog Creek

A memoir of origin, sovereignty, and lived refusal.

Following the Compass

On movement, orientation, and learning to trust direction.

The Fixed Point

Stillness, identity, and what remains when motion stops.

Learning & Formation

The Color of Nothing

Lessons on learning, attention, and the space between.

Outside the Bell

A book about children whose intelligence does not ring on cue.

The Bird Who Looked First

A story about early seeing.

The Capsule and the World

How intelligence forms before the world tells it what to be.

Work, Systems & Aftermath

The Art After the Deal

What remains when the transaction is over.

When the Forms Stopped Coming

Short stories from a post–income-tax world.

When Money Changed Quietly

On invisible systems and everyday life.

The First Answer Trap

On thinking with AI instead of stopping at the first reply.

Home, Attention & Everyday Life

The Kitchen Without a Recipe

On cooking, intuition, ingredients, and the intelligence of making.

Relational & Perceptual Works

Clarity, Sovereignty & Signal

Operating Systems & Structural Works

The Freedom Stack & Cash Flow is Oxygen
A system for building real financial freedom.

Structure Without Walls
On execution, fluid systems, and what holds without rigidity.

The Marriage Operating System
Why love fails without structure.

The Style Operating System
How style becomes consistent through structure.

The Corporate Structure
What org charts hide and systems reveal.

The Creative OS
Where attention becomes structure, and structure protects the signal.

The Aging OS
Where aging becomes structure, and structure protects aliveness.

The Judgment Operating System
Where judgment remains human.

The Constitutional Operating System
How a free people govern power.

The Party-State Operating System
How power organizes itself to endure.

The Impossible Operating System
What SpaceX reveals about how impossible becomes inevitable.

The Human Advantage Operating System
Where human capacity becomes visible again.

Capstone

The Wayfinder Maps
An integration of the whole.

www.lisajscottpublishing.com

LISA J. SCOTT
PUBLISHING